AF413260

WHAT ARE
TENSORS
EXACTLY?

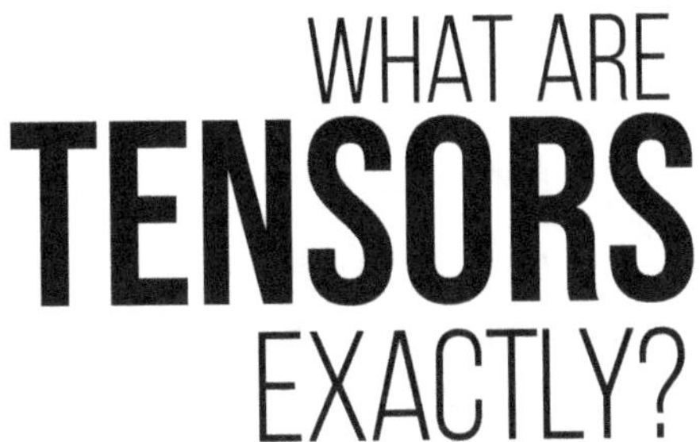

WHAT ARE TENSORS EXACTLY?

HONGYU GUO

UNIVERSITY OF HOUSTON-VICTORIA, USA

World Scientific

NEW JERSEY · LONDON · SINGAPORE · BEIJING · SHANGHAI · HONG KONG · TAIPEI · CHENNAI · TOKYO

Published by

World Scientific Publishing Co. Pte. Ltd.

5 Toh Tuck Link, Singapore 596224

USA office: 27 Warren Street, Suite 401-402, Hackensack, NJ 07601

UK office: 57 Shelton Street, Covent Garden, London WC2H 9HE

British Library Cataloguing-in-Publication Data
A catalogue record for this book is available from the British Library.

WHAT ARE TENSORS EXACTLY?

ISBN 978-981-124-101-7 (hardcover)
ISBN 978-981-124-102-4 (ebook for institutions)
ISBN 978-981-124-103-1 (ebook for individuals)

For any available supplementary material, please visit
https://www.worldscientific.com/worldscibooks/10.1142/12388#t=suppl

To Yanping and Alicia

Preface

Tensors have profound applications in physics and engineering. There is often a fuzzy haze surrounding the concept of tensor when it is defined in the old-fashioned way using the component approach. The tensor is defined as a matrix, but amended with the transformation laws. It is defined as the components of an object, without a clear definition of what this object is. It gives an impression of an equivocal duality of matrix and non-matrix, just like the mixture of the living and the dead states of Schrödinger's cat. What especially confuses students is the coexistence of the old and the new definitions in literature. The appearances of these definitions look so different that students can hardly guess that they are referring to the same thing. The old-fashioned definition is difficult to understand because it is not rigorous; the modern definitions are difficult to understand because they are rigorous but at a cost of being more abstract and less intuitive. It is the goal of this book to elucidate the rigorous definitions of tensor in an intuitive way, so that students no longer have to recite those definitions like a parrot.

The audience of this book is graduate students, higher level undergraduate students, as well as researchers and professionals in physics and engineering. The book can also benefit students of mathematics major to build more intuition about tensors. The prerequisite to this book is basic linear algebra. Some concepts of linear algebra are reviewed right before they are used. More advanced topics in linear algebra, like covectors and dual spaces, contravariant and covariant components of vectors, bilinear forms and quadratic forms are supplied in Appendix 1. The point of view of mathematical structures advocated by Bourbaki is very helpful in studying modern mathematics, including tensors and Riemannian geometry. Readers unfamiliar with this can find it in Appendix 2.

Chapter 1 is an introduction and overview of tensors. Chapter 3 is a short chapter about direct sum spaces. Chapter 4 through Chapter 7 are mainly tensor algebra. Chapter 2, Chapter 8 and Chapter 9 discuss applications in machine learning and physics. Chapter 10, the last chapter, provides an outlook on Riemannian geometry and general relativity (see chapter dependency chart after the Table of Contents). More advanced topics which are out of the scope of this book are marked with an asterisk in front of the section title. The boxes include remarks which are excursions away from the main logical thread of the subject, most of which are historical notes and the philosophical views of my own.

Acknowledgments

The following images are from Wikimedia Commons under the Creative Commons license: Figure 1.3 by Thomas Schultz; Figure 1.4, a snapshot of a 3D reconstruction by P. Hagmann et al.; Figure 2.1 by Wesalius; Figure 2.2 by Zinskauf; Figure 10.7 by Strebe, modified. I would like to give my sincere gratitude to these authors.

I am deeply indebted to Profs. Ricardo Teixeira, Jerry Hu and Ali Dogan, and my graduate students Vu Pham and Kapil Suryawanshi at the University of Houston—Victoria. They took their precious time reviewing the draft manuscript and finding errors. I am also grateful to Profs. Guangming Xing and Zhonghang Xia at the Western Kentucky University for reading part of the manuscript and giving me feedback. I would like to thank the staff of World Scientific Publishing, especially Steven Patt, the desk editor, for the assistance in the production of this book. My special thanks go to Rajesh Babu, the deputy manager of production, for his assistance on technical issues while I was typesetting the text in LaTeX. He is that kind of macho TeX programmer described by Leslie Lamport. I am deeply impressed by his capability of solving all sort of hard problems. Last but foremost, I would like to thank two beautiful and loving ladies, my wife and my daughter, for their constant support.

Houston, May 2021

Hongyu Guo
guoh@uhv.edu

Contents

Chapter 7. Tensor Algebra

81

Chapter 8. Dynamics: The Inertia Tensor

85

Chapter 9. Electrodynamics: The EM Field Tensor

103

Chapter 10. Riemannian Geometry and General Relativity

119

Appendix 1. Topics of Linear Algebra

179

Appendix 2. Mathematical Structures   193

Appendix 3. Axiomatic Systems   199

List of Boxes

List of Figures

Chapter Dependency Chart

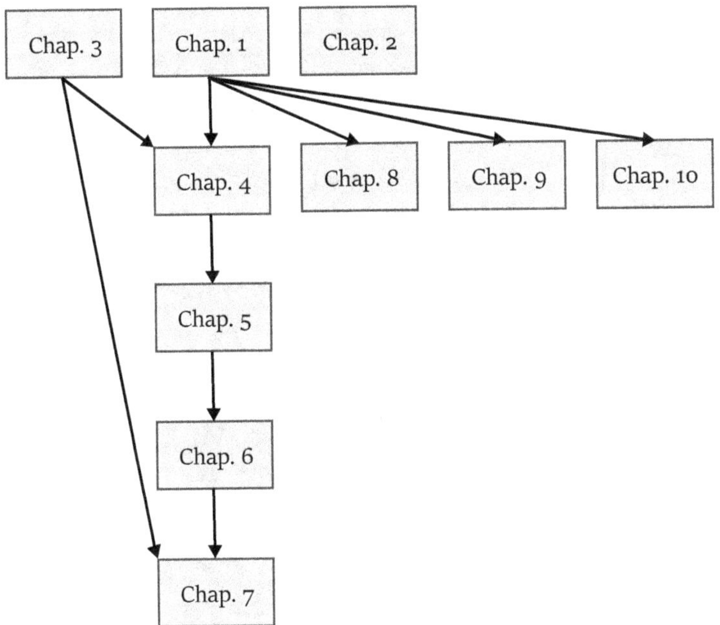

Notation

Notation	Meaning
$\overset{\text{def}}{=}$	equal by definition
$\forall$	for all
$\exists$	there exists
$\exists!$	there exists unique
$\circ$	mapping composition
$(+) : U \times V \to W$	$(u, v) \mapsto u + v$; infix operator $+$ is put in ().
$() : U \times V \to W$	$(u, v) \mapsto uv$; infix operator is omitted.
$\mathbb{R}$	the set of real numbers
$\mathbb{C}$	the set of complex numbers
$\mathbb{R}^n$	n-dimensional real vector space
$\mathbb{E}^n$	n-dimensional Euclidean space
$\operatorname{Im}\varphi$	image of mapping φ
$\det A$	determinant of matrix A
$\operatorname{rank}A$	rank of matrix A
A^t	transpose of matrix A
$\dim V$	dimension of V
$\langle S \rangle$ or $\operatorname{Span}(S)$	linear subspace spanned by set S
$\langle \mathbf{u}, \mathbf{v} \rangle$ or $\mathbf{u} \cdot \mathbf{v}$	inner product or dot product
$\oplus$	direct sum
$\otimes$	tensor product
$V^{\otimes(p,q)}$ or $T_q^p(V)$	tensor space of type (p, q) over V
$\Pi^{\circ\circ}_{\quad\downarrow} : T_0^2(V) \to T_1^1(V)$	lowering one index of the tensor
$\Pi^{\circ\circ}_{\downarrow\downarrow} : T_0^2(V) \to T_2^0(V)$	lowering two indices of the tensor
$\Pi^{\circ\uparrow}_{\circ} : T_1^1(V) \to T_0^2(V)$	raising one index of the tensor
$\Pi^{\uparrow\uparrow}_{\circ\circ} : T_2^0(V) \to T_0^2(V)$	raising two indices of the tensor
∂_x	partial derivative, shorthand for $\frac{\partial}{\partial x}$
$\wedge$	exterior product or wedge product
$d\wedge$	exterior derivative

Chapter 1

Confusions: What Are Tensors Exactly?

One way to learn a lot of mathematics is by reading the first chapters of many books.

— Paul R. Halmos

Tensors have profound applications in physics, computer science, engineering, machine learning, data mining, medicine (diffusion tensor imaging), etc. This chapter provides a background overview of tensors. You may find usage of terms that have not yet been defined. The purpose is to have a "big picture".

If you find the first chapter helpful, you might consider reading beyond it. The logical exposition starts in Chap. 3.

§1. Questions and Confusions

The concept of tensor is confusing to many students. If one does a search on the Internet, he can find many questions asked about tensors. For example:

> Is a tensor just a (higher dimensional) matrix?
> How long have tensors been around, and why is there a sudden fascination for tensors in machine learning?
> Are tensors in machine learning the same thing as tensors in mathematics and physics?
> Are tensors in machine learning contravariant or covariant?
> What is a metric tensor?
> Why is inertia tensor a tensor? (It is defined as a matrix in most of the books.)
> What is an example of a quantity that has the correct number of components but fails to be a tensor?
> What is the connection between tensor and tensor product?
> What is the physical meaning of a tensor?
> Can you add the components of a contravariant tensor and a covariant tensor?
> Do pure mathematicians have an interest in tensor analysis?
> What are some open problems in tensor analysis?
> Is tensor analysis relevant to deep learning?

There are many answers and explanations floating on the Internet. However, instead of solving the mysteries, many of these only add more confusion to the already confused learners. The following are a few examples:

> "A tensor is just an n-dimensional array with n indices."
> "Tensors are simply mathematical objects that can be used to describe physical properties."
> "Tensors are generalizations of scalars and vectors."
> "Basically tensors are vectors which have not a single direction but they rather point in all directions."
> "If I ask you what a vector is, you may tell me that is an element of a vector space, so tensor is an element of a tensor space."

"Tensors have properties of both vectors and scalars, like area, stress etc."

"A tensor is not a scalar, a vector or anything. It's just an abstract quantity that obeys the coordinate transformation law. Anything that satisfies the law is a tensor. That's it!"

"In mathematics, tensors are geometrical objects that describe the linear relationships between geometric, numerical, and other tensile vectors."

"The simplest way to imagine a tensor is that it's a vector in a product space. Each index denotes a factor of the product space in which the tensor lives, and may be raised or lowered depending on how the corresponding factor transforms under a change of basis. The number of indices counts the rank of a tensor. As such, tensors are essentially just generalizations of vectors. Their components (in a certain basis) are multidimensional arrays. A tensor is more than simply a multidimensional array, for the same reason that a vector is not simply a list of its components."

"Speaking somewhat non-technically, tensors represent a linear operator of other tensors. Each time you operate a tensor on another tensor a set of matching indices disappears."

"A tensor is a multilinear function."

"A tensor, with the possibility of a multitude of indices, both covariant and contravariant, look like multidimensional data in 0, 1, 2, 3, and higher dimensions."

"In the simplest form: the quantity having magnitude, direction and plane to act are called tensor quantities."

"A tensor is an element of a tensor product of two or more vector spaces."

"A tensor is the tensor product of two vectors."

"Tensor: it is those physical quantity which may have tension-like effects."

Well, each of them speaks some truth about tensors, but they also reflect a lot of confusions. This reminds me of reading some funny answers of young children to the question "What is love".

*** Comparison:** What do love and tensor have in common?

"What is love?"

> "Love is when a girl puts on perfume and a boy puts on shaving cologne and they go out and smell each other." (age 5)
>
> "Love is when you tell a guy you like his shirt, then he wears it every day." (age 7)
>
> "If you want to learn to love better, you should start with a friend who you hate." (age 6)
>
> "Love is when mommy sees daddy smelly and sweaty and still says he is handsomer than Robert Redford." (age 8)
>
> "Love is when your puppy licks your face even after you left him alone all day." (age 4)
>
> "Love is when you kiss all the time. Then when you get tired of kissing, you still want to be together and you talk more." (age 8)
>
> "I know my older sister loves me because she gives me all her old clothes and has to go out and buy new ones." (age 4)
>
> "I let my big sister pick on me because my mom says she only picks on me because she loves me. So I pick on my baby sister because I love her." (age 4)

Each of these answers certainly tells some aspect of the truth.

What do love and tensor have in common? Is the love between sisters the same as that between mom and dad, dating teenagers, and dogs and humans? Compare with the question: is the tensor in machine learning the same as those in mathematics and physics?

The concept of love is abstract and complex, and it has never been rigorously defined. The tensor is also abstract and complex. It was poorly defined in the past. There are rigorous modern definitions, but at a cost of being more abstract and less intuitive. So the old-fashioned definition is hard to understand because it is not rigorous; the modern definition is hard to understand because it is rigorous. It is the goal of this book to explain the rigorous definitions of tensor in an intuitive way, so that students no longer have to recite those definitions like a parrot.

We shall have answers to these questions through this book. After reading the book, the reader should be able to judge the above quoted answers, which is correct and which is wrong. However, readers would like to have some quick answers before committing to reading a book. That is the purpose of this chapter.

§2. Who Invented the Tensor?

In this section, we give a brief history of the concept of tensor. This answers the question how long tensors have been around. It also answers the question "why are tensors confusing" from one perspective: it has different origins and it is the merge of different threads in history. In the next section we provide answers to this question from another aspect: there are many apparently different definitions of tensor in the current literature.

There were several threads in the development of tensor theory in late 1800s and early 1900s, including Ricci, Gibbs, Voigt and Whitney. Most modern authors give credit to Ricci for the concept of tensor, because the early textbooks, especially the physics literature, predominantly followed his definitions. Ricci did not use "tensor" in his definition, but rather "system". Physicists transplanted the name "tensor" to Ricci's definition. Although being called a "tensor", Ricci's definition actually defines a tensor field. This causes the most confusion to the beginners. Gibbs, Voigt and Whitney defined a tensor as a tensor in the algebraic sense.

(1) G. Ricci [(1892)]: covariant and contravariant systems, but he called those "systems", rather than "tensors" (what he defined is a tensor field in the modern sense; see more in Sec. 3).

(2) J. W. Gibbs [(1884)]: dyadics and polyadics (these are actually tensors in the modern sense, only by different names; see more in Chap. 4).

(3) W. Voigt [(1898)]: coined the name tensor—in a narrower sense of symmetric tensors in the study of elasticity of crystals.

(4) H. Whitney [(1937)]: tensor product (see more in Chap. 5).

Gibbs is recognized as one of the founders of vector algebra and vector analysis. Gibbs played an important role in emancipating vectors from Hamilton's quaternions. What is often underappreciated is his major contribution in the development of tensor algebra and tensor analysis (in

Euclidean space). Gibbs developed the concept of dyadics and polyadics. These are actually tensors in the modern sense, only by different names.[1] His dyadic product is exactly the tensor product in the modern sense, except his notation is the juxtaposition of two vectors $\mathbf{uv}$, compared with the modern notation of $\mathbf{u} \otimes \mathbf{v}$.

W. Voigt [(1898)] introduced the term tensor, in his study of stress and strain of crystals in his book *The Fundamental Physical Properties of the Crystals* (*Die fundamentalen physikalischen Eigenschaften der Krystallen*). The word "tensor" has its root "tensus" in Latin, meaning stretch or tension. Both stress and strain tensors are symmetric tensors of the second order and each has six components. Voigt denotes them as a 6-dimensional vector. This is known as the Voigt notation. The term tensor was adopted by physicists Max Abraham (1904), Arnold Sommerfeld (1910), Max von Laue (1911). Einstein and Grossmann [(1913)] [2] used Ricci's definition but with the name "tensor" instead of Ricci's name "system".

Whitney [(1937)] defined the tensor product. It is actually the idea of Gibbs dyadics made more precise. There are also other threads that are related to the development of tensors. Grassmann developed exterior algebra in 1862. Although exterior algebra can be established independent of the tensor theory, there is a connection between these two. An exterior vector is in fact an antisymmetric tensor. H. Minkowski [(1908)] introduced the electromagnetic tensor, which is an antisymmetric tensor, although he called it a "vector of the second kind" (of 6 dimensions, to distinguish it from a "vector of the first kind" with 4 dimensions). A. Sommerfeld later called it a 6-vector. Let us compare it with Voigt's tensor for stress, which is also expressed as a 6-vector. Voigt's tensor is a symmetric tensor over a 3-dimensional vector space, while the electromagnetic field tensor is an antisymmetric tensor over a 4-dimensional vector space.

Chap. 9 discusses the electromagnetic field tensor.

---

[1] The term tensor did appear in Gibbs' book, but was used to refer to a special type of tensors (namely a special type of linear transformations). W. R. Hamilton also used the term tensor, but referring to the modulus of a quaternion, which is totally irrelevant to our tensor theory. Tensor in Hamilton's sense is no longer in use today. Rather, it is called the modulus or norm of the quaternion.

[2] This paper has two parts put together, with Einstein as the single author for the physics part and Grossmann as the single author for the mathematics part.

*** Philosophical View**: Is mathematics invented or discovered?
—My opinion: It is both.

We asked the question "who invented the tensor". Was the tensor invented, or discovered? There is even an age-long philosophical question: "Is mathematics invented, or discovered?"

We asked the question "what is a tensor". In fact, a tensor is whatever we define it to be. We do have the liberty when it comes to definitions. In this sense, mathematics is an invention. Sherman Stein [(2010)] wrote a book, *Mathematics: the Man-made Universe*. The title of the book reflects this view. Of course, other people have argued that mathematics is discovery and this topic has been an unresolved debate.

My opinion is: it is both. In mathematics, we first invent this man-made universe. Then we make discoveries inside it. This man-made universe can be extremely complex and discovery in it is by no means a trivial process. For instance, the creation of non-Euclidean geometry is an invention, but its interpretations (or models) are discoveries, which uncover the connection between non-Euclidean and Euclidean geometries. Take group as another example. The definition of a group takes only a few lines of text, which can be viewed as an invention. The culminating result in group theory, the classification of the finite simple groups is a discovery, with tens of thousands of pages in several hundred articles written by about 100 authors, published mostly between 1955 and 2004. Riemannian manifold can be another example. Its definition also consists of just a few lines of text. The Nash embedding theorem is a great discovery, which reveals that although Riemannian manifold is defined intrinsically, it is always isometric to some submanifold embedded in some higher dimensional Euclidean space.

I have interpreted discovery as the discovery in the man-made universe of mathematics itself. Is mathematics about discovery in nature? My answer is yes and no: *no* in the sense that modern mathematics in its abstract form is liberated from the obligation of discovering the truth in nature, but *yes* in the sense that mathematics may be part of the process of discovering nature when it is applied in science. In the old days, mathematics was intended to discover the truth in nature directly, but in modern days, its participation in the discovery is indirect. Whatever abstract mathematics can be applied to the real world, if we find a *physical model* of the abstract *mathematical structure* (Appendix 2).

§3. Different Definitions of the Tensor

Why is the concept of tensor confusing? It is just a definition, isn't it? Think about the definition of an equilateral triangle. No one would have difficulty with that.

Some factors may make a concept hard to understand:

(1) The concept itself is more complex.

(2) The definition itself is not clear. Oftentimes the lack of rigor in the definition is caused by the intrinsic complexity of the concept itself. Historically, the first attempts to define a concept were often not successful in pinning down the essence of the concept. It may take centuries for the concept to evolve and get crystallized. Mathematics is full of evolution history of such concepts: complex numbers, real numbers, limit, continuity, vectors, ..., and the list goes on and on (see the boxes at the end of the section).

(3) Different definitions coexist in the literature, also due to historical reasons. Some of these definitions are equivalent, but not all of them are equivalent.

It turns out that all these factors have an effect on the concept of tensor. They cause many confusions for the beginners. In the following, we list several definitions of tensors that can be found in textbooks. Don't worry if you are confused with these. It is just to show that you do have a good reason to be confused, which is not your fault.

Definitions 1 and 2 are mostly seen in older textbooks of tensor analysis, physics, and especially general relativity.

Definition 1. A set of quantities ξ^{rs} is said to be a **contravariant tensor** (of degree 2) if under the change of coordinates

$$x'^i = x'^i(x^1, \ldots, x^n), \quad i = 1, \ldots, n, \tag{1.1}$$

they transform according to

$$(\xi')^{st} = \sum_{\sigma, \tau} \xi^{\sigma\tau} \frac{\partial x'^s}{\partial x^\sigma} \frac{\partial x'^t}{\partial x^\tau}. \tag{1.2}$$

A set of quantities ξ_{lm} is said to be a **covariant tensor** if they transform according to

$$(\xi')_{lm} = \sum_{\lambda, \mu} \xi_{\lambda\mu} \frac{\partial x^\lambda}{\partial x'^l} \frac{\partial x^\mu}{\partial x'^m}. \tag{1.3}$$

A set of quantities $\xi_l{}^s$ is said to be a **mixed tensor** if they transform according to

$$(\xi')_l{}^s = \sum_{\lambda,\sigma} \xi_\lambda{}^\sigma \frac{\partial x^\lambda}{\partial x'^l} \frac{\partial x'^s}{\partial x^\sigma}. \qquad (1.4)$$

Remark. This definition is basically due to Ricci. It is confusing that most books call these tensors, but what Ricci defines here are actually tensor fields. Ricci should not be blamed because he called these "systems". It is the use of the name tensor [Einstein and Grossmann (1913)] that causes the confusion of tensors with tensor fields. Each "quantity", or component ξ^{rs} is actually a function of space locations $\mathbf{x} = (x_1, \ldots, x_n)$. If the set of quantities is considered a single tensor ξ, then Ricci defines a tensor field $\xi(\mathbf{x})$, which is the assignment of a tensor ξ to each space point $\mathbf{x}$. A tensor ξ should be a single algebraic entity. Logically, a tensor as an algebraic entity should be defined first, before the definition of a tensor field, but this was not done by Ricci. This is the reason why Ricci used the components in his definition but amended by the coordinate transformation laws. In the modern perspective, these transformation laws are not necessary. They are the consequence of the basis change in the tangent space of the differentiable manifold, induced by local coordinate change Eq. 1.1 (see Sec. 3 in Chap. 10).

The arbitrary coordinate transformation Eq. 1.1 and the involvement of partial derivatives in the above definition clearly hint the tensor field. To make a seemingly algebraic definition of tensor, the general coordinate transformation Eq. 1.1 is restricted to linear transformations. This results in the following shy version of the definition.

Definition 2. A set of quantities ξ^{rs} is said to be a **contravariant tensor** (of degree 2) if under the change of coordinates

$$x'^i = \sum_k \Lambda_k{}^i x^k \qquad (1.5)$$

and its inverse

$$x^k = \sum_i \bar{\Lambda}_i{}^k x'^i, \qquad (1.6)$$

where the constant coefficients $\Lambda_k{}^i$ and $\bar{\Lambda}_i{}^k$ satisfy

$$\sum_r \Lambda_i{}^r \bar{\Lambda}_r{}^k = \delta_i{}^k, \qquad (1.7)$$

they transform according to

$$(\xi')^{st} = \sum_{\sigma,\tau} \xi^{\sigma\tau} \Lambda_\sigma{}^s \Lambda_\tau{}^t. \tag{1.8}$$

A set of quantities ξ_{lm} is said to be a **covariant tensor** if they transform according to

$$(\xi')_{lm} = \sum_{\lambda,\mu} \xi_{\lambda\mu} \bar\Lambda_l{}^\lambda \bar\Lambda_m{}^\mu. \tag{1.9}$$

A set of quantities $\xi_l{}^s$ is said to be a **mixed tensor** if they transform according to

$$(\xi')_l{}^s = \sum_{\lambda,\sigma} \xi_\lambda{}^\sigma \bar\Lambda_l{}^\lambda \Lambda_\sigma{}^s. \tag{1.10}$$

Remark. Although this version looks more algebraic, the meaning of the linear coordinate transformation Eq. 1.5 is still not clear, if the set of quantities is an individual tensor instead of a tensor field. Furthermore, the meanings of "contravariant" and "covariant" are not apparent. According to K. Reich [(1994)], J. Sylvester introduced the terms "covariant" and "contravariant" in 1851 [Sylvester (1851)]. We shall reveal this in Sec. 2 of Chap. 6, these coordinate changes are with respect to the basis change of the underlying vector space, which involves a matrix $A_i{}^k$. Eq. 1.7 tells us that $\bar\Lambda_i{}^k$ is the transpose of the inverse of $\Lambda_k{}^i$. The matrix $\bar\Lambda_i{}^k$ here is same as $A_i{}^k$ in Sec. 2 of Chap. 6. That is why the transformation of covariant tensor involves $\bar\Lambda_i{}^k$, which means "the same as", or "together with" the transformation of the basis, while the contravariant tensor involves $\Lambda_k{}^i$, which is the inverse of $A_i{}^k$ with a meaning "against". We may call the basis transformation the "forward" transformation and its inverse the "backward" transformation. If the basis undergoes a forward transformation, the coordinates will undergo a "backward" transformation, as in Eq. 1.5, with an analogy: if the train moves forward, the trees outside seem to move in the backward direction from the perspective of someone inside the train. So the transformation for contravariant tensors is really "contra" to the basis transformation, which is not explicit here. It is rather "together with" the coordinate transformation of vectors Eq. 1.5. Eq. 1.5 itself is considered "contra", or "backward", with respect to the basis transformation. Another word of caution for the beginners is the popular tensor component notation in literature. Although $\bar\Lambda$ looks similar to Λ, it is actually the transpose of the inverse matrix of Λ.

g^{ij} are the components of the inverse matrix of the metric matrix g_{ij}.

This kind of definition of tensor is often referred to as the old-fashioned definition. It is this component approach that caused the conundrum, with the concept of tensor portrayed as an equivocal duality of matrix and non-matrix, just like the mixture of the living and the dead states of Schrödinger's cat. The tensor is defined as a matrix, but amended by the transformation laws. It is defined as the components of an object, without a clear definition of what this object is.

In recent years, with the booming research in machine learning, the machine learning community uses the tensor simply in the sense of a multi-dimensional array (or higher dimensional matrix), ignoring the transformation laws and breaking up this fuzzy duality. We shall discuss tensors in machine learning in Chap. 2.

Definition 3. (in the context of machine learning) A tensor is a multi-dimensional array (or matrix).

It is a trend in recent physics textbooks to use the following definition of a tensor.

Definition 4. Let V be a vector space over $\mathbb{R}$ and V^* be its dual space. A multilinear mapping

$$\Phi : \underbrace{V^* \times \cdots \times V^*}_{p} \times \underbrace{V \times \cdots \times V}_{q} \to \mathbb{R}$$

is called a tensor of type (p, q).

Remark. A question from a curious student arises naturally. In this definition, why does the co-domain of the multilinear mapping Φ have to be the real numbers $\mathbb{R}$? Can $\mathbb{R}$ be replaced by some other vector space? Is a multilinear mapping $\Psi : V \times \ldots \times V \to V$ a tensor? In particular, is a linear transformation $\varphi : V \to V$ a tensor?

The answer to these questions is that this definition is only a model of tensors. A cat is an example (model) of animals, while not all the animals are cats. There are other models of tensors which are not covered in this definition. We shall show (see more in Sec. 8 of Chap. 5) that indeed a multilinear mapping $\Psi : V \times \ldots \times V \to V$ is a vector-valued tensor. In particular, a linear transformation $\varphi : V \to V$ is a tensor. A quadratic form $\phi : V \to \mathbb{R}$ is also a tensor (quadratic forms are closely related to bilinear forms; see Appendix 1).

The following defines a tensor space (tensor product space). Then an element of this space is called a tensor. This is the abstract approach, and this is what we are going to adopt in the main course of this book (see Chap. 5).

Definition 5. (Tensor product space) Let U, V and W be vector spaces, and $\otimes : U \times V \to W$ be a bilinear mapping. The pair $(W, \otimes)$ is called a **tensor product space** (or simply **tensor space**) over the underlying vector spaces U and V, if they satisfy the following conditions:
(1) Generating property

$$W = \langle \mathrm{Im}\otimes \rangle \, ;$$

(2) Maximal span property

$$\dim W = \dim U \cdot \dim V.$$

The vectors in W are called **tensors** over U and V. The mapping $\otimes$ is called the **tensor multiplication** of two vectors, or **tensor product mapping**, or simply **tensor product**, or **tensor mapping**. W is often denoted by $U \otimes V$.

Remark. The coordinate change laws in the old-fashioned definition are only the phenomena. The essence of tensors is the multilinearity, or multilinear mappings. The coordinate change laws are the consequences of the multilinear mapping—tensor product mapping. In history, the multilinearity was understood by Gibbs and Ricci but was not emphasized explicitly.

The following definition is often seen in textbooks in pure mathematics.

Definition 6. Let U, V and W be vector spaces and suppose $\otimes : U \times V \to W$ is a bilinear mapping. $(W, \otimes)$ is called a **tensor product space** of U and V if the following conditions are satisfied (unique factorization property):
For any vector space X and any bilinear mapping $\Psi : U \times V \to X$, there exists a *unique* linear mapping $\varphi : W \to X$ such that

$$\Psi = \varphi \circ \otimes.$$

Remark. Some authors prefer this definition because it is terse in language, and it applies not only when U and V are finite dimensional spaces, but also when they are infinite dimensional vector spaces. It is not a good choice as a definition from the perspective of pedagogy for beginners. We shall treat this as a theorem about the universal property after the tensor product space is defined in an alternative way.

The following definition is based on construction (see the *Encyclopedic Dictionary of Mathematics* [Mathematical Society of Japan (1993)]; see also [Bourbaki (1942); Roman (2005)]). It describes the intuitive ideas of Gibbs dyadics but it is made rigorous in modern abstract language.

Definition 7. Let U and V be vector spaces over the same field F. Let $\mathscr{V}_F \langle U \times V \rangle$ be the free vector space generated by $U \times V$. Let Z be the subspace of $\mathscr{V}_F \langle U \times V \rangle$ generated by all the elements of the form

$$a(\mathbf{u}_1, \mathbf{v}) + b(\mathbf{u}_2, \mathbf{v}) - (a\mathbf{u}_1 + b\mathbf{u}_2, \mathbf{v}),$$
$$a(\mathbf{u}, \mathbf{v}_1) + b(\mathbf{u}, \mathbf{v}_2) - (\mathbf{u}, a\mathbf{v}_1 + b\mathbf{v}_2),$$

for all $a, b \in F$, $\mathbf{u}, \mathbf{u}_1, \mathbf{u}_2 \in U$ and $\mathbf{v}, \mathbf{v}_1, \mathbf{v}_2 \in V$.

The quotient space

$$U \otimes V = \frac{\mathscr{V}_F \langle U \times V \rangle}{Z}$$

is called the **tensor product** of U and V. The elements in $U \otimes V$ are called **tensors** over U and V.

Define a mapping $\otimes : U \times V \to U \otimes V$ such that for all $\mathbf{u} \in U$ and $\mathbf{v} \in V$, $(\mathbf{u}, \mathbf{v}) \mapsto \mathbf{u} \otimes \mathbf{v} \overset{\text{def}}{=} [(\mathbf{u}, \mathbf{v})]$, where $[(\mathbf{u}, \mathbf{v})]$ is the equivalence class of $(\mathbf{u}, \mathbf{v})$ in $\mathscr{V}_F \langle U \times V \rangle$ defined by the subspace Z. This mapping is a bilinear mapping and is called the **canonical bilinear mapping**.

We have listed many different definitions of the tensor, which are commonly seen in textbooks. All of these are not exactly equivalent (some of them do, in some sense), but rather they reflect the historical evolution of the tensor concept.

* **Historical Note**: Evolution of definitions in mathematics

Many mathematical concepts are complex and difficult in nature. These concepts were not crystal clear when they were initially invented. These concepts have an evolutionary history and the definitions have been refined through time. Such examples are abundant, such as complex numbers, irrational numbers, real numbers, vectors, length, area, volume, probability, function, continuous function, Dirac delta function, infinity, infinitesimal, set, etc. Tensor is just one more example which can be added to the list. There have been occasions when a mathematician defined a new concept, it was even difficult for his contemporary fellow mathematicians to understand. Take Grassmann's exterior algebra for example. Heinrich Baltzer wrote to August Möbius after reading Grassmann's book *Ausdehnungslehre*: "It is not now possible for me to enter into those thoughts; I become dizzy and see sky-blue before my eyes when I read them." Möbius replied: "If as you write me, you have not relished Grassmann's *Ausdehnungslehre*, I reply that I have the same experience. I likewise have managed to get through no more than the first two sheets of his book."

* **Historical Note**: What are vectors exactly?

The concept of vector has gone through a similar long history of evolution as well. Some physical quantities like velocity and force are quantities with a magnitude and a direction. The parallelogram law of vector addition was known in Newton's time but the name vector was not used. The name vector was coined by Hamilton to denote the imaginary part $bi + cj + dk$ of his quaternion $a + bi + cj + dk$. It was Gibbs and Heaviside who liberated the vector from the shackles of the quaternion and made it an independent entity. At that time, vectors were mainly confined to three dimensions. This was soon generalized to higher dimensions and a vector was defined as an n-tuple. It was Peano who defined the vector space in the abstract sense in 1888. However, he did not use the name vector space, or linear space, but rather he called it a "linear system". (Interestingly, compare with the history of tensors. Ricci did not use the name "tensor", but rather a "system" instead.) Look at the following definitions of a vector.

(1) A vector is a quantity with a magnitude and a direction.

(2) A vector is an n-tuple of numbers.

(3) A vector is an element in a vector space.

These are not exactly equivalent definitions, but rather they reflect the historical evolution of the concept. Definition (2) is in terms of components. Definition (3) is abstract and axiomatic. With the definitions (2) and (3), a vector does not automatically have a magnitude.

A high school student often learns (1) as the definition of a vector in a physics course, but (2) as the definition in a mathematics course. He is likely to be confused with the question: are the vectors in physics and mathematics the same thing? The confusion shall be cleared when they learn the abstract definition of vector space in college, because (1) and (2) are just models of the abstract vectors.

The history of tensors is along a similar line. In this book, we are going to study the abstract, or axiomatic definition, and relate different concrete models to it.

*** Historical Note**: What are imaginary numbers exactly?

The typical definition of complex number in high school textbooks is: A complex number is *a number* that can be written in the form $a + bi$, where a and b are real numbers and i is the imaginary unit defined by $i^2 = -1$. This definition follows Jerome Cardan, who conceived it in 1545 without a solid logical foundation. The concept then kept evolving in the next three centuries to come, going through the initial confusion and denial to the final clarification and acceptance. Cardan himself considered these numbers as "mental tortures" and "useless". Descartes coined the term "imaginary" and rejected it. It was Gauss who named it "complex number" to rescue it from the mystery of the "imaginary" domain. Even Euler made a mistake in writing $\sqrt{-1}\sqrt{-4} = \sqrt{4} = 2$ in his book *Algebra*. It is a paradoxical argument by applying $\sqrt{a}\sqrt{b} = \sqrt{ab}$ to obtain $\sqrt{-1}\sqrt{-1} = \sqrt{(-1)(-1)} = 1$ (or similarly, $i^2 = (\sqrt{-1})^2 = \sqrt{(-1)^2} = 1$).

The geometrical representation due to Argand marked a big step toward demystifying imaginary numbers. The modern definition of complex number is due to Hamilton in 1837: A complex number is an ordered pair (a, b) of real numbers. The number $(a, 0)$ is identified with the real number a, and i is defined as the pair $(0, 1)$. The addition and multiplication of complex numbers are defined by

$$(a_1, b_1) + (a_2, b_2) \stackrel{\text{def}}{=} (a_1 + a_2,\ b_1 + b_2),$$

$$(a_1, b_1) \cdot (a_2, b_2) \stackrel{\text{def}}{=} (a_1 a_2 - b_1 b_2,\ a_1 b_2 + a_2 b_1).$$

By this definition, $i^2 = (0, 1) \cdot (0, 1) = (-1, 0) = -1$.

*** Historical Note**: What are irrational numbers exactly?

This is basically the same question as "what are the real numbers exactly", because an irrational number can be defined as a real number that is not a rational number. Rational numbers are easier to define. The essence of a rational number is the ratio of two integers. A rational number can be defined as the equivalence class of a pair of integers. To many people's surprise, the concept of real numbers is much more complex than complex numbers. Logically, the concept of real numbers should precede that of complex numbers because a complex number is defined as a pair of real numbers, but historically, the rigorous definition of real numbers came much later than that of complex numbers. The concept of irrational numbers emerged from incommensurable segments in ancient Greek geometry and was used intensively in the early development of calculus without a rigorous definition. The rigorous definitions of real numbers, like Dedekind cuts and Cantor's construction through Cauchy sequences, finally came in the nineteenth century. In this sense, the complex number $\sqrt{-1}$ is much simpler than $\sqrt{2}$, because the latter involves infinite sets.

*** Historical Note**: What are sets exactly?

Georg Cantor was the founder of set theory, which serves as the foundation of modern mathematics. The concept of set, as a collection of objects, is intuitive. However, it is not precise. For example, we could think of a set U, which is the set of all sets. Since U is also a set, it is a member of itself—$U \in U$. There are other sets x with the property $x \notin x$. This leads to the Russell's paradox. Let us construct a set $Q \stackrel{\text{def}}{=} \{x | x \notin x\}$. Now we ask the question: is Q a member of itself?

Namely, is $Q \in Q$ true? First, suppose $Q \in Q$. Then Q does not satisfy the property $x \notin x$, and hence $Q \notin Q$. Next, suppose $Q \notin Q$. Then Q satisfies the property $x \notin x$. Hence $Q \in Q$. A popular version of this is the barber paradox: a barber in a village, who is a man, claims that he shaves every man in the village who does not shave himself, and does not shave any man who shaves himself. Now there is a question: does the barber shave himself? According to his claim, he shaves himself if and only if he does not shave himself.

Gottlob Frege was a German logician, who made significant contributions in logic. Russell's paradox was a big blow to him. He became depressed and did no serious mathematics thereafter. Unlike physicists (see Sec. 6 of Chap. 10; see also [Guo (2021)][3]), mathematicians take paradoxes seriously. What is a way out of this paradox? It is actually pretty simple. We redefine the concept of set more precisely so that those trouble makers like U and Q no longer qualify to be called sets. It is not an ordinary definition. The qualification is regulated by a set of axioms introduced by Zermelo and Fraenkel. These axioms are actually the hidden definition of set (see more on axiomatic systems in Appendix 3).

§4. Plain Things by Fancy Tensor Names

Quite some terms bear the surname "Tensor", like metric tensor, curvature tensor, inertia tensor, stress tensor, diffusion tensor imaging, etc. These are just fancy names for plain things, which may sound intimidating to beginners. Yes, they are tensors and it is not wrong to call them tensors, but tensor theory is not essential to understand these concepts. They can go by other names without the use of "tensor". Calling them tensors is like calling water by the name "dihydrogen monoxide". Everyone understands water, but people may be confused by the chemistry jargon.

These terms were named historically because of the fact that they are (represented by) matrices. The confusion is rooted in the question whether a tensor is the same as a matrix. If it does, why don't we simply call them metric matrix, inertia matrix, etc.? The old-fashioned definition of tensor is equivocal about whether a tensor is simply a matrix or not. A tensor is defined as a matrix of components, but amended awkwardly by the transformation laws.

---

[3]Guo, H. (2021). A New Paradox and the Reconciliation of Lorentz and Galilean Transformations, *Synthese*, https://doi.org/10.1007/s11229-021-03155-y (open access).

Things get clear with the modern view. The metric tensor is just an inner product, the inertia tensor can be defined as a linear transformation or a quadratic form. The stress tensor and diffusion tensor are simply linear transformations. We shall discuss inertial tensor in more detail in Chap. 8, and the metric tensor for Riemannian geometry in Chap. 10.

Think of the stress forces in liquids and solids. In a liquid, let us single out a small piece of imaginary surface, which separates the liquid on both sides. Each side exerts a force on the other side (Figure 1.1a). Let us use a vector $\mathbf{S}$ to represent the surface, where $\mathbf{S}$ is a normal vector of the surface, and the magnitude of $\mathbf{S}$ represents the area of the surface. Let $\mathbf{F}$ be the vector representing the force that the liquid on one side exerts on the other side. Because liquids cannot have shear forces, the force $\mathbf{F}$ must be in the normal direction of the surface, which is the same as $\mathbf{S}$. $\mathbf{F}$ is linearly related to $\mathbf{S}$,

$$\mathbf{F} = \sigma\mathbf{S}, \tag{1.11}$$

where σ is a scalar coefficient, which is called the pressure.

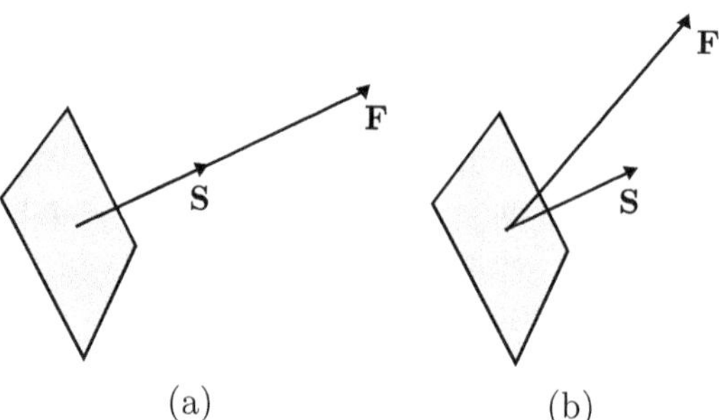

Figure 1.1   (a) Stress in liquids (b) Stress in solids

Things are different in solids, like crystals. The force $\mathbf{F}$ in general is not in the same direction as $\mathbf{S}$. $\mathbf{F}$ can be decomposed into normal stress, and shear stress (in the tangent direction of the surface). However, $\mathbf{F}$ is still linearly related to $\mathbf{S}$ (Figure 1.1b). This relation is a linear transformation:

$$\mathbf{F} = \Sigma\mathbf{S}, \tag{1.12}$$

where Σ is a linear transformation which can be represented by a matrix $[\Sigma]$ with components σ_{ij},

$$\begin{bmatrix} F_1 \\ F_2 \\ F_3 \end{bmatrix} = \begin{bmatrix} \sigma_{11} & \sigma_{12} & \sigma_{13} \\ \sigma_{21} & \sigma_{22} & \sigma_{23} \\ \sigma_{31} & \sigma_{32} & \sigma_{33} \end{bmatrix} \begin{bmatrix} S_1 \\ S_2 \\ S_3 \end{bmatrix}.$$

Σ is called the stress tensor. This can be written as

$$F_i = \sum_{j=1}^{3} \sigma_{ij} S_j. \tag{1.13}$$

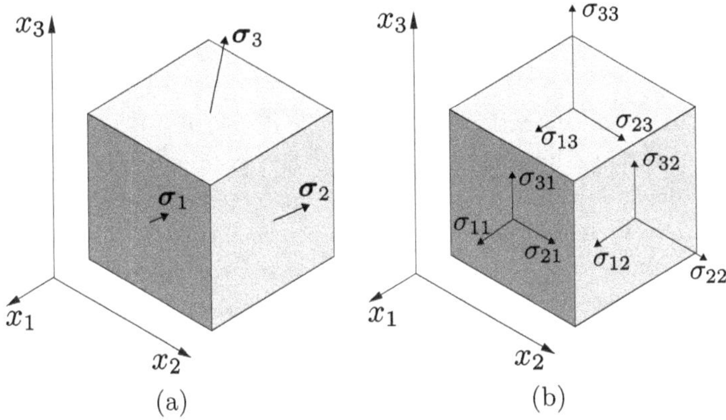

Figure 1.2   (a) Stress tensor as three vectors (b) The nine components of the stress tensor

The matrix of the stress tensor Σ can be viewed as three column vectors

$$\boldsymbol{\sigma}_1 = \begin{bmatrix} \sigma_{11} \\ \sigma_{21} \\ \sigma_{31} \end{bmatrix}, \quad \boldsymbol{\sigma}_2 = \begin{bmatrix} \sigma_{12} \\ \sigma_{22} \\ \sigma_{32} \end{bmatrix}, \quad \boldsymbol{\sigma}_3 = \begin{bmatrix} \sigma_{13} \\ \sigma_{23} \\ \sigma_{33} \end{bmatrix}.$$

What are the physical meanings of these three vectors? Imagine we have a small cube. Their faces are along the three axes with normal vectors $\mathbf{s}_1 = (1,0,0)$, $\mathbf{s}_2 = (0,1,0)$, $\mathbf{s}_3 = (0,0,1)$ and unit area. $\boldsymbol{\sigma}_1$ is the stress force acted on the face $\mathbf{s}_1$, $\boldsymbol{\sigma}_2$ is the stress force acted on the face $\mathbf{s}_2$, and so on (Figure 1.2a). Each force $\boldsymbol{\sigma}_i$ has three components and together the stress matrix has nine components. What is the physical meaning of the component σ_{ij}? σ_{ij} represents the ith component of $\boldsymbol{\sigma}_j$, which is the force acting on the face $\mathbf{s}_j$ (orthogonal to x_j axis). On face $\mathbf{s}_1$, σ_{11} is the normal stress while σ_{21} and σ_{31} are the tangent stresses. On face $\mathbf{s}_2$, σ_{22} is the normal stress while σ_{12} and σ_{32} are the tangent stresses (Figure 1.2).

In fact, the tensor here is just a linear transformation, and the stress tensor Σ is just one example of linear transformations used in physics. Eq. 1.13 is the component form of any linear transformation, not just limited to the stress situation. The linear transformation maps any vector $\mathbf{S}$ to a new vector $\mathbf{F} = \Sigma\mathbf{S}$, as in Eq. 1.12. The meaning of its component σ_{ij} is the ith component of $\mathbf{F}$ when $\mathbf{S}$ is a unit vector along the jth direction. Here we have given a physical interpretation of the linear transformation Σ in the example of stress in solids, or crystals.

The physical process of diffusion in isotropic media is described by Fick's law:

$$\mathbf{J} = -d\nabla\phi,$$

where ϕ is the concentration density of the diffusive substance, which is a function of the spatial location $\mathbf{x}$; $\nabla\phi$ is the gradient of ϕ; $\mathbf{J}$ is the flux of the diffusive substance, and d is a scalar constant called the diffusion coefficient. However, in anisotropic media, the flux $\mathbf{J}$ is usually not in the same direction as $\nabla\phi$, but it still has a linear relationship with $\nabla\phi$. This means that $\mathbf{J}$ and $\nabla\phi$ are related by a linear transformation:

$$\mathbf{J} = -D\nabla\phi.$$

This linear transformation D is often called the diffusion tensor and it has nine components when a coordinate system is chosen. In coordinate form, it can be written as

$$J_i = -\sum_{j=1}^{3} D_{ij}\frac{\partial\phi}{\partial x_j}.$$

The brain consists of gray matter and white matter. The gray matter consists of the neuron bodies while the white matter consists of the myelinated axon fibers, which serve as the interconnections between the neurons. The diffusion of water in the brain is highly anisotropic due to these axon fibers. With the help of magnetic resonance imaging (MRI), the diffusion tensor components at space locations can be measured, which is used to reconstruct the fiber tracts in the brain. This is known as diffusion tensor imaging (DTI). Figure 1.3 shows the diffusion tensor field (represented by ellipsoids, see Sec. 5 of Chap. 8). Figure 1.4 shows the reconstructed fiber tracts of the brain using DTI.

Figure 1.3   Diffusion Tensor Imaging: ellipsoids of the diffusion tensors

Figure 1.4   Diffusion Tensor Imaging: fiber tracks in the brain white matter

§5. Tensors without a Tensor Name— Linear Transformations

Many objects that we are familiar with are actually tensors, but they do not often go by a tensor name. We shall show that linear mappings and linear transformations are tensors. Realizing these mundane objects are actually tensors has a demystifying effect. Here is just the gospel. The details will be discussed in Chaps. 5 and 6.

When a basis of the vector space V is chosen, a linear transformation $\varphi : V \to V$ can be represented by a matrix. When the basis is changed, the matrix of the linear transformation changes in accordance. This explains why the tensors in the old-fashioned definition have to obey the transformation laws, and most importantly, it explains what causes the transformations.

Suppose $\langle \cdot, \cdot \rangle$ is an inner product defined in V. Given two constant vectors $\mathbf{a}, \mathbf{b} \in V$, we define a linear transformation:

$$\varphi_{\mathbf{a},\mathbf{b}} : V \to V;$$

$$\mathbf{x} \mapsto \varphi_{\mathbf{a},\mathbf{b}}(\mathbf{x}) \stackrel{\text{def}}{=} \mathbf{a} \langle \mathbf{b}, \mathbf{x} \rangle , \text{ for all } \mathbf{x} \in V.$$

Basically, the vector $\mathbf{x}$ is projected onto $\mathbf{b}$ and the inner product $\langle \mathbf{b}, \mathbf{x} \rangle$ is calculated. The final output is a vector along the direction of $\mathbf{a}$ but scaled by the factor $\langle \mathbf{b}, \mathbf{x} \rangle$.

The vector $\mathbf{b}$ here can be viewed as a linear function in the dual space V^*. The effect of $\mathbf{b}$ acting on a vector $\mathbf{x} \in V$ is $\mathbf{b}(\mathbf{x}) = \langle \mathbf{b}, \mathbf{x} \rangle$. The linear transformation $\varphi_{\mathbf{a},\mathbf{b}}$ is actually the tensor product in $V \otimes V^*$ and we denote $\varphi_{\mathbf{a},\mathbf{b}} = \mathbf{a} \otimes \mathbf{b}$.

A beginner might be tempted to guess that all the linear transformations can be put in the form of $\mathbf{a} \otimes \mathbf{b}$, for some $\mathbf{a} \in V$ and $\mathbf{b} \in V^*$, but this is not true. However, any linear transformation can be written as the sum of these tensor products, $\mathbf{a}_1 \otimes \mathbf{b}_1 + \ldots + \mathbf{a}_k \otimes \mathbf{b}_k$. Therefore, a linear transformation is a mixed tensor of type $(1, 1)$, and of course, it obeys the transformation law in Eq. 1.4. This is also why the inertia tensor, stress tensor and diffusion tensor are tensors, but using plain words, they are just linear transformations.

A linear transformation is also a special case of a more general model—vector-valued tensor, which is a multilinear mapping $\Phi : V_1 \times \ldots \times V_q \to X$. When $q = 1$ and $V_1 = X = V$, we have a linear transformation $\Phi : V \to V$. We discuss vector-valued tensors in Sec. 8 of Chap. 5.

§6. Comparison: Different Definitions of the Vector
—Concrete Systems vs. Abstract Systems

To better understand the concept of tensor, we make a comparison with the vector, which we are already familiar with. The key to understand the difficulty associated with tensors is the appreciation of the relationship between the abstract concepts and concrete examples.

Historically, there have been different definitions of vectors too. These definitions are not exactly equivalent and they reflect the historical evolution of the concept.

Definition 8. A vector is a quantity with a magnitude and a direction.

Definition 9. A vector is a directed line segment in space. The addition of two vectors is defined by the parallelogram law.

Definition 10. A vector is an n-tuple of real numbers $(x_1, \ldots, x_n)$.

Definition 11. Let F be a field and V a nonempty set. V together with two operations called addition $(+) : V \times V \to V$ and scalar-vector multiplication $() : F \times V \to V$, is called a vector space over F, if these operations satisfy the following conditions. The elements in V are called vectors and the elements of F are called scalars.

(1) $(\mathbf{u} + \mathbf{v}) + \mathbf{w} = \mathbf{u} + (\mathbf{v} + \mathbf{w})$.
(2) There exists $\mathbf{0} \in V$ such that $\mathbf{u} + \mathbf{0} = \mathbf{u}$.
(3) For any $\mathbf{u} \in V$, there exists $\mathbf{x} \in V$ such that $\mathbf{u} + \mathbf{x} = \mathbf{0}$. We denote $\mathbf{x} = -\mathbf{u}$.
(4) $a(\mathbf{u} + \mathbf{v}) = a\mathbf{u} + a\mathbf{v}$.
(5) $(a + b)\mathbf{u} = a\mathbf{u} + b\mathbf{u}$.
(6) $a(b\mathbf{u}) = (ab)\mathbf{u}$.
(7) $1\mathbf{u} = \mathbf{u}$, where $1 \in F$ is the multiplicative identity in F.

A reader may have already learned that the vector space is an Abelian (commutative) group with respect to the vector addition, but finds that the commutative law $\mathbf{u} + \mathbf{v} = \mathbf{v} + \mathbf{u}$ is missing from the above list of axioms. These axioms were first proposed by Peano. He included this commutative law and almost all the textbooks afterwards just followed him. However, this axiom is not independent of the rest, and hence there is no need to list it explicitly (see a proof in Appendix 1). Peano was a master with the axiomatic systems. It is remarkable that he devised this axiomatic system for vector space (which he called linear system) as early as 1888. Amazingly all of the axioms, except the commutative law of addition, turned out to be independent.

Remark. Definition 8 is traditional and vague. Definition 10 is more general than Definition 9, as it defines an n-dimensional vector while the vector in Definition 9 is 3-dimensional.

Definition 11 is the most general and the most abstract of all. It is an axiomatic definition. Any system that satisfies these axioms is called a model of the abstract vector space. Vectors defined in Definitions 9 and 10 are examples, or models of a vector space. We can find many other models of vectors in the following.

Example 1. (Matrix spaces) All $m \times n$ real matrices $M_{m,n}$ form a real vector space with respect to matrix addition and matrix multiplication by a number. Each $m \times n$ matrix is a vector.

Example 2. (Linear mappings) Let V and W be vector spaces. All linear mappings $\varphi : V \to W$ form a vector space. Each linear mapping is a vector.

Example 3. (Polynomials of degree at most n) All polynomials with real coefficients of degree at most n, form a real vector space with respect to polynomial addition and multiplication by a number. Each polynomial is a vector.

Example 4. (All polynomials) All polynomials of one variable with real coefficients form a real vector space with respect to addition and multiplication by a number. Each polynomial is a vector. This vector space is infinite dimensional.

Example 5. (Real functions) All real functions $f : \mathbb{R} \to \mathbb{R}$ form a real vector space. If f, g are two real functions and $a, x \in \mathbb{R}$, we define $f + g = h$, where $h(x) = f(x) + g(x)$; and $(af)(x) = af(x)$. Each real function is a vector. This vector space is infinite dimensional.

Despite the large number of apparently different models, there is one interesting property. That is, any model of an n-dimensional vector space is isomorphic to each other, in particular, isomorphic to the vector space of

n-tuples in Definition 10. Because of this isomorphism, we have the liberty of choosing the abstract Definition 11, or the concrete Definition 10.

The different definitions of tensors also reflect the history of evolution of the concept.

Definition 5 for tensors is in a similar position to Definition 11 for vectors. It is an abstract or axiomatic definition. Definitions 3, 4 and 7 are models of the abstract tensor.

§7. Tensor Product and Tensor Spaces

We can ask two different but related questions:

"What is a tensor?"

"What is a tensor space?"

Definitions 3 and 4 define an individual tensor, while Definition 5 defines an abstract tensor (product) space $U \otimes V$, and any element in this space is called a tensor.

We shall discuss tensor product spaces in Chap. 5 and tensor power spaces $V^{\otimes p} = V \otimes \ldots \otimes V$ in Chap. 6.

When we talk about tensor spaces $U \otimes V$ or $V^{\otimes p}$, we should not neglect the relationship between the tensor space $V^{\otimes p}$ and the vector space V. We call V the *underlying vector space* of tensor space $V^{\otimes p}$.

There is a good comparison with vector spaces. Recall, in a vector space, there are two distinct sets, the set of vectors V and the set of scalars, which is a field F. V is called the "vector space *over* the field F" and F is called the ground field of V (Figure 1.5).

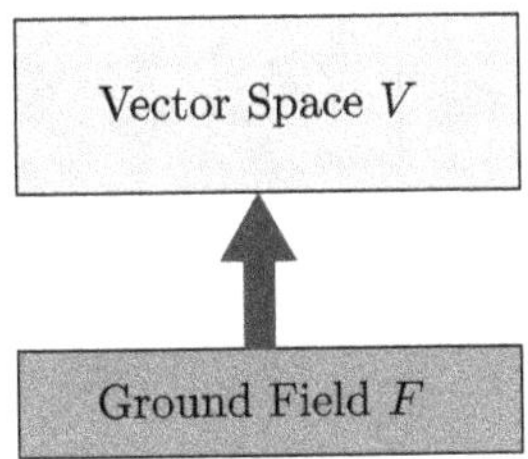

Figure 1.5   Vector space V and its ground field F

The interaction between the ground field F and vector space V is through the scalar-vector multiplication $() : F \times V \to V$.

The relationship between the underlying vector space and the tensor space is the tensor product, which is a bilinear mapping $\otimes : V \times V \to V \otimes V$. From this point of view, the tensor space $V^{\otimes 2} = V \otimes V$ is a vector space by itself. A tensor is also a vector. This view is different from the traditional view that tensors are generalizations of vectors because their transformation laws are different (Figure 1.6).

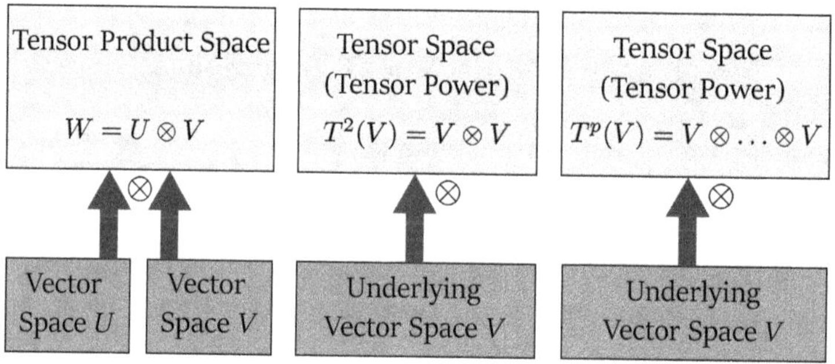

Figure 1.6   Tensor space $V^{\otimes p}$ and its underlying vector space V

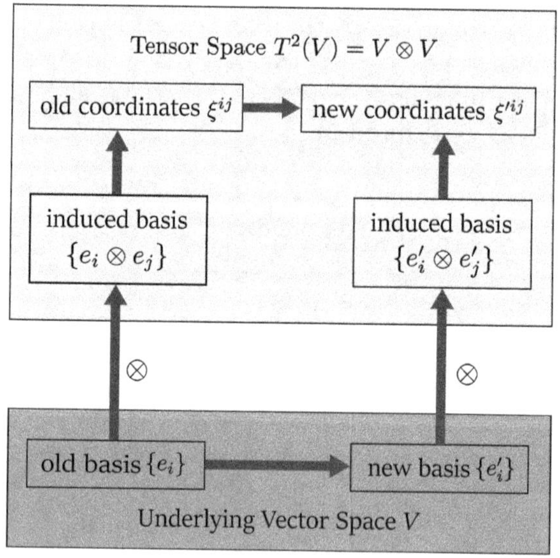

Figure 1.7   Coordinate change of a tensor

Given a basis $\{\mathbf{e}_1,\ldots,\mathbf{e}_n\}$ for V, the tensors $\{\boldsymbol{\tau}_{ij}|\boldsymbol{\tau}_{ij}=\mathbf{e}_i\otimes\mathbf{e}_j,\, i,j=1,\ldots,n\}$ form a basis for the tensor space $V^{\otimes 2}$, which contains n^2 basis vectors. When the basis $\{\mathbf{e}_1,\ldots,\mathbf{e}_n\}$ of V changes, the induced basis $\{\boldsymbol{\tau}_{ij}\}$ for $V^{\otimes 2}$ changes to $\{\boldsymbol{\tau}'_{ij}\}$ accordingly. Then the change of coordinates of a tensor in $V^{\otimes 2}$ obeys those laws in Definition 2. Therefore those coordinate change laws refer to coordinate changes of the tensors in $V^{\otimes 2}$ in response to the basis change of V, rather than in response to the basis change of $V^{\otimes 2}$ itself, which is also a vector space (Figure 1.7). Therefore, a tensor is also a vector, rather than a generalization of a vector. We could use a single index running from 1 to n^2 for the tensor components. If its basis changes, the components of a tensor in $V^{\otimes 2}$ with a single index will just behave like a vector (Figure 1.8). The reason we adopt double indices ij is the relationship between V and $V^{\otimes 2}$, which is the tensor product $\otimes$.

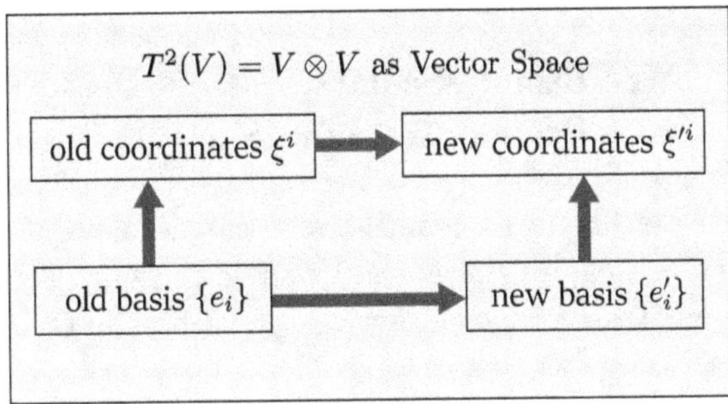

Figure 1.8   Coordinate change of a tensor as a vector

§8. Degree, Rank, Order or Dimension— Which Is the Best Name?

One may encounter a mixture of terms in literature—rank, order and degree, used interchangeably. They all mean the same thing, the number of indices of a tensor component. In the machine learning community, they even use "dimension" for this, because they use the term tensor as a multidimensional array.

Ricci never used the term "tensor" in his writings. He called it a "system". He also used the term "order" of a system. Physicists use "rank" more often.

However, in the modern view, the tensor space $V^{\otimes p} = V \otimes \ldots \otimes V$

is the p-th tensor power (tensor product of the same vector space with itself p times). It is natural to call p the degree, drawing similarity with the naming of the degree of polynomials. This naming agrees with the *Encyclopedic Dictionary of Mathematics* [Japanese Mathematical Society (1993)], which is an excellent reference source and provides the standard terminology of modern mathematics.

Following N. Bourbaki [(1942)], the term rank of a tensor is defined with a different meaning from the degree. Recall the rank of a square matrix (similarly for a linear transformation) is defined as the number of linearly independent columns (or rows). An $n \times n$ square matrix can have any rank between 1 and n. A tensor of degree 2 may have any rank between 1 and n. Any decomposable tensor of degree 2 has a rank of 1 (see more in Sec. 5 of Chap. 5).

*§9. What Are Pseudo-Scalars, Pseudo-Vectors and Pseudo-Tensors Exactly?

In older physics textbooks, some authors introduce the concepts of pseudo-scalars, pseudo-vectors, and in general pseudo-tensors. They are also defined by different transformation laws. Let us first look at the so called pseudo-vectors.

This is the definition: a quantity is called a pseudo-vector (or axial vector) if it transforms like a vector under proper transformation (for example, rotation), but the transformation gains an additional sign flip under an improper transformation.

A proper transformation reserves the orientation of an oriented vector space while an improper transformation changes the orientation. For example, the reflection $x' = -x$, $y' = -y$, $z' = -z$ is an improper transformation.

One example of a pseudo-vector is illustrated as the cross product $\mathbf{w} = \mathbf{u} \times \mathbf{v}$. They argue, for a regular vector (also called polar vector), when the coordinates go through a reflection, $\mathbf{v}$ should be transformed to $\mathbf{v}' = -\mathbf{v}$. But for the cross product, $\mathbf{w}' = (-\mathbf{u}) \times (-\mathbf{v}) = \mathbf{w}$. Magnetic field and angular momentum are examples of pseudo-vectors.

One example of a pseudo-scalar is the triple scalar product (representing signed volume) of three vectors $a = \mathbf{v}_1 \cdot (\mathbf{v}_2 \times \mathbf{v}_3)$. When the coordinates go through a reflection, $a' = (-\mathbf{v}_1) \cdot [(-\mathbf{v}_2) \times (-\mathbf{v}_3)] = -a$.

This argument does not seem to make sense. A scalar is just a number and it should not depend on coordinates. Why should it be affected by coordinate reflection and change sign accordingly?

A closer examination reveals that something is not expressed clearly and logically in these concepts. We take the pseudo-vector for example. Let V and W be 3-dimensional vector spaces, $\mathbf{u}, \mathbf{v} \in V$ and $\mathbf{w} \in W$. V and W are isomorphic, but let us distinguish them. Now we view the cross product as a mapping $(\times) : V \times V \to W$. Here $\times$ is not a tensor product mapping, but it is a bilinear mapping in a similar situation. It connects spaces V and W. Let $\mathbf{w} = \mathbf{u} \times \mathbf{v}$, and let $\{\mathbf{b}_1, \mathbf{b}_2, \mathbf{b}_3\}$ be a basis for V. We define $\mathbf{e}_1, \mathbf{e}_2, \mathbf{e}_3 \in W$,

$$
\begin{aligned}
\mathbf{e}_1 &\stackrel{\text{def}}{=} \mathbf{b}_2 \times \mathbf{b}_3, \\
\mathbf{e}_2 &\stackrel{\text{def}}{=} \mathbf{b}_3 \times \mathbf{b}_1, \\
\mathbf{e}_3 &\stackrel{\text{def}}{=} \mathbf{b}_1 \times \mathbf{b}_2.
\end{aligned}
\tag{1.14}
$$

Then $\{\mathbf{e}_1, \mathbf{e}_2, \mathbf{e}_3\}$ forms a basis for W. After coordinate reflection, the new induced basis vectors are

$$
\begin{aligned}
\mathbf{e}'_1 &\stackrel{\text{def}}{=} \mathbf{b}'_2 \times \mathbf{b}'_3 = (-\mathbf{b}_2) \times (-\mathbf{b}_3) = \mathbf{e}_1, \\
\mathbf{e}'_2 &\stackrel{\text{def}}{=} \mathbf{b}'_3 \times \mathbf{b}'_1 = (-\mathbf{b}_3) \times (-\mathbf{b}_1) = \mathbf{e}_2, \\
\mathbf{e}'_3 &\stackrel{\text{def}}{=} \mathbf{b}'_1 \times \mathbf{b}'_2 = (-\mathbf{b}_1) \times (-\mathbf{b}_1) = \mathbf{e}_3.
\end{aligned}
$$

Therefore, $\mathbf{w}$ has the same coordinates under induced basis $\{\mathbf{e}'_1, \mathbf{e}'_2, \mathbf{e}'_3\}$ as under basis $\{\mathbf{e}_1, \mathbf{e}_2, \mathbf{e}_3\}$. This is also explained with Figure 1.7 in a similar way, except now the mapping is the cross product $\times$, instead of the tensor product $\otimes$. This means, as a 3-tuple and a member of W, $\mathbf{w}$ is certainly an ordinary vector. If the space W is unrelated to V, when the basis of W goes through a reflection, the coordinates of $\mathbf{w}$ with respect to the new basis of W certainly flip the sign. When we say $\mathbf{w}$ is a pseudo-vector and the signs of $\mathbf{w}$ do not change, we are talking with respect to the induced basis $\mathbf{e}'_1 = \mathbf{b}'_2 \times \mathbf{b}'_3$, $\mathbf{e}'_2 = \mathbf{b}'_3 \times \mathbf{b}'_1$ and $\mathbf{e}'_3 = \mathbf{b}'_1 \times \mathbf{b}'_2$, which are induced by the cross product.

After all, the pseudo-vectors can be viewed as living in a vector space W. The pseudo-vectors are just ordinary vectors and transform as ordinary vectors with respect to a basis change in W itself. However, there is a connection between the vector space W with another underlying vector space V. In general, let us denote it by $\circledast : V \times V \to W$. The coordinates of a pseudo-vector in W changes like a pseudo-vector with respect to basis change in V composed with the mapping $\circledast$.

The cross product only applies in 3-dimensional vector spaces. For the general n-dimensional vector space V, the pseudo-vectors can be viewed as living in the space of $\Lambda^{n-1}(V)$, which is the exterior space over V to the $(n-1)$-th power. It has the same dimension as V. The pseudo-vector in $\Lambda^{n-1}(V)$ can be viewed as the Hodge dual of a vector in V. A pseudo-scalars can be viewed as living in the space of $\Lambda^n(V)$, which is the dual of $\Lambda^0(V) \overset{\text{def}}{=} \mathbb{R}$ and has dimension 1.

For a pseudo-tensor of degree two, it transforms as

$$(\xi')^{st} = \text{sign}(\Lambda) \sum_{\sigma,\tau} \xi^{\sigma\tau} \Lambda_\sigma{}^s \Lambda_\tau{}^t,$$

where $\text{sign}(\Lambda)$ is the sign of $\det \Lambda$. This extra sign can also be viewed as the result of some bilinear mapping connecting the space of pseudo-tensors W to the underlying vector space V,

$$\circledast : V \times V \to W.$$

The more general concept is the tensor density of weight k, with a transformation law

$$(\xi')^{st} = (\det \Lambda)^k \sum_{\sigma,\tau} \xi^{\sigma\tau} \Lambda_\sigma{}^s \Lambda_\tau{}^t,$$

where $\det \Lambda$ is the determinant of the transformation matrix Λ in the underlying vector space V, and k is a constant exponent.

§10. What Is Tensor Analysis Exactly? Relation to Riemannian Geometry

10.1   Vector Analysis

Vector analysis studies vector-valued functions. Let V be a vector space over $\mathbb{R}$. A vector-valued function can be a function of a single variable $\mathbf{p} : \mathbb{R} \to V; t \mapsto \mathbf{p}(t)$, or a function of multiple variables, like $\mathbf{f} : \mathbb{R}^3 \to V; (x, y, z) \mapsto \mathbf{f}(x, y, z)$. $\mathbf{p}(t)$ is often interpreted as a vector which changes with time t, while $\mathbf{f}(x, y, z)$ is a vector field, with a vector $\mathbf{f}$ assigned to each spatial location (x, y, z). So vector analysis is the differential calculus of vector fields, while the single variable vector functions can be viewed as a special case.

Gibbs was a pioneer of vector analysis. His book [Gibbs (1884)] deals with both vector algebra and vector analysis. In vector analysis, three differential operators on vector (or scalar) fields are defined: the gradient of a scalar field $\nabla\varphi$, the divergence of a vector field $\nabla\cdot\mathbf{f}$ and the curl (or rot, for rotation) of a vector field $\nabla\times\mathbf{f}$. Important theorems involving these operators include Gauss' theorem

$$\iiint_V (\nabla\cdot\mathbf{f})dV = \oiint_{\partial V} \mathbf{f}\cdot d\mathbf{S},$$

Stoke's theorem

$$\iint_S (\nabla\times\mathbf{f})\cdot d\mathbf{S} = \oint_{\partial S} \mathbf{f}\cdot d\mathbf{r},$$

and properties like

$$\nabla\times(\nabla\varphi) = 0,$$
$$\nabla\cdot(\nabla\times\mathbf{f}) = 0.$$

10.2   Tensor Analysis and Riemannian Geometry

Some people view tensors as the generalization of vectors, and it is natural to guess that the study of tensors should be divided into tensor algebra and tensor analysis, with the latter studying the differential calculus of tensor fields in Euclidean space $\mathbb{R}^3$. As a matter of fact, tensor analysis in this sense was also developed by Gibbs in his book of vector analysis. Gibbs used different terminology but his dyadics and polyadics are just tensors in the modern sense. He defined several algebraic operations—dot products and cross products for dyads, which can be linearly extended to general tensors, like

$$\mathbf{a}\cdot(\mathbf{bc}) \overset{\text{def}}{=} (\mathbf{a}\cdot\mathbf{b})\mathbf{c},$$
$$(\mathbf{ab})\cdot\mathbf{c} \overset{\text{def}}{=} \mathbf{a}(\mathbf{b}\cdot\mathbf{c})$$
$$(\mathbf{ab})\cdot(\mathbf{cd}) \overset{\text{def}}{=} (\mathbf{b}\cdot\mathbf{c})\mathbf{ad},$$
$$\mathbf{a}\times(\mathbf{bc}) \overset{\text{def}}{=} (\mathbf{a}\times\mathbf{b})\mathbf{c},$$
$$(\mathbf{ab})\times\mathbf{c} \overset{\text{def}}{=} \mathbf{a}(\mathbf{b}\times\mathbf{c}),$$
$$(\mathbf{ab}):(\mathbf{cd}) \overset{\text{def}}{=} (\mathbf{a}\cdot\mathbf{d})(\mathbf{b}\cdot\mathbf{c}),$$

etc. Along this line, viewing the nabla operator ∇ as a vector operator, the gradient of a vector $\nabla\mathbf{u}$, the gradient, divergence and curl of tensors $\nabla(\mathbf{uv})$, $\nabla\cdot(\mathbf{uv})$, $\nabla\times(\mathbf{uv})$ and many other operations can be defined. Gibbs did explore the properties of these operations and demonstrated many applications in physics and mathematics, including applications to the curvature of surfaces in differential geometry.

However, tensor analysis in this direction of studying tensor fields in Euclidean space $\mathbb{R}^3$ has not gone too far in history, because it is kind of trivial. What is called tensor analysis today is in the context of Riemannian geometry. The tensor fields are assumed to be tensor fields on a Riemannian manifold, or a differentiable manifold in general.

Ricci called his work absolute calculus, with an emphasis on the covariant derivative. Levi-Civita contributed the concept of parallel transport. Levi-Civita did not use the term tensor in his early works, but adopted this new name in his book [Levi-Civita (1927)] *The Absolute Differential Calculus (Calculus of Tensors)* after Einstein and Grossmann had popularized the term tensor.

However, tensor analysis is not really a new branch, or independent branch of mathematics. It is just Riemannian geometry in a slightly different dialect, characterized by the component (or index) form of representation. In his *Mathematical Thought from Ancient to Modern Times*, M. Kline [(1972)] writes:

> "Tensor analysis is often described as a totally new branch of mathematics, created *ab initio* either to meet some specific objective or just to delight mathematicians. It is actually no more than a variant on an old theme, namely, the study of differential invariants associated primarily with a Riemannian geometry."

The "differential invariant associated primarily with a Riemannian geometry" that Kline refers to is the fundamental form $ds^2 = \sum_{i=1}^{n} g_{ij}dx_idx_j$, or the line element, or the metric tensor, which is the higher dimensional generalization of Gauss' first fundamental form. It is invariant under coordinate transformations (or isometric mappings, in the active view), or re-parameterizations (the passive view). The characteristic of Ricci's absolute differential calculus, or tensor analysis is the component approach. É. Cartan [(2002)] recommended, "as far as possible avoid very formal computations in which an orgy of tensor indices hides a geometric picture which is often very simple." Chap. 10 provides an outlook of Riemannian geometry and general relativity but it is not the scope of this book to go deeper than that. The reader is referred to [Bishop and Goldberg (1980)] and [Guo (2014)] for further reading.

Chapter 2

Why and How Are Tensors Used in Machine Learning?

In recent years there has been a boom in machine learning research, especially deep learning, marked by the victory of the machine program AlphaGo over the best human Go players, the popularization of Google TensorFlow application framework and the Google customized tensor processing unit (TPU) for machine learning applications. Tensor has been adopted as the basic data structure for machine learning applications. We do not intend to have a deep discussion of machine learning topics, but we shall address the role of tensor in machine learning briefly.

§1. How AlphaGo Beat the Best Human Go Player via Deep Learning

In March 2016, a computer Go program AlphaGo developed by Google's DeepMind shocked the world. It beat the 18-time world champion Lee Sedol by 4 to 1 in a five-game match [Silver et al. (2016)]. Go, or the siege game, is an ancient Chinese board game, which has been played continuously for more than 2,500 years to the present day. It enjoys great popularity in the CJK cultures (China, Japan and Korea). The name in Chinese is *weiqi*, with *wei* meaning siege or surrounding and *qi* meaning board game. In Japanese, the Chinese characters *weiqi* are pronounced *igo*, with *i* for *wei* and *go* for *qi*. Its rules are very simple: two players take turns to place black and white stones on a 19×19 grid. Whoever occupies a bigger territory wins. If a group of stones are completely surrounded by the opponent's stones, the opponent takes over the surrounded territory.

The victory of AlphaGo in 2016 is a great milestone in game AI. A previous milestone was the match in 1997 when the IBM's chess program Deep Blue beat the resigning world champion Garry Kasparov. Compared with chess, Go has a much higher complexity. The number of possible board positions of Go is approximately 2.1×10^{170}, much greater than the estimated number of atoms 10^{80} in the observable universe. Go has an average branching factor of 250, and depth of the search tree 150, while chess has an average branching factor of 35 and a depth of 80. Moreover, people believed that Go was such a type of game that favored human intuition over machine's brute-force computation.

A core element of AlphaGo's algorithm is Monte Carlo tree search, which is a random search method, proposed by Chang et al. [(2005)] and improved by Coulom [(2006)]. The idea of random search is simple and intuitive and it makes sense when the game tree is big (with big branching factors and depth). According to Chang, "in practice, most implementations of Monte Carlo tree search, including all of those in the best Go-playing computer programs, use an algorithm called UCT (upper confidence bound 1 applied to trees) introduced in Kocsis and Szepesvári [(2006)], based on the UCB1 formula of Auer et al. [(2002)] and the provably convergent algorithm first applied to multi-stage decision-making models (specifically, Markov decision processes) by Chang et al. [(2005)]."

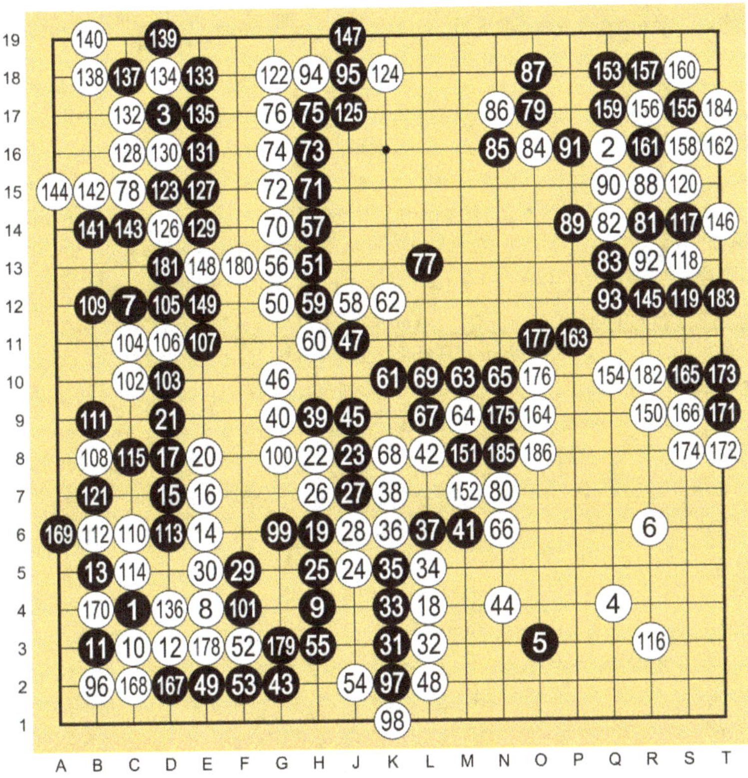

Figure 2.1   Game 1: AlphaGo (white) vs. Lee Sedol (black), 2016; AlphaGo wins.

Monte Carlo tree search is an improvement and can be applied in various games. It led to strong performance in chess, checkers and Othello, but weak amateur level play in Go. The key to further improvement is a smart random search. That is, we do not search the game tree with uniform randomness, but rather, we give priorities in the subtrees that are more promising. The decision of priority should be guided by prior experience. This is the concept of learning. To do this, AlphaGo implements two deep neural networks, a policy network, and a value network, used to narrow down the search. These are used as factors that assign priorities (with probabilities) to maintain a balance of *exploration* and *exploitation*.

*** Philosophical View**: What is intelligence exactly?
—My opinion: high complexity

There has never been a satisfactory definition of intelligence. Why? Oftentimes, when we make a definition, it is not just a definition. There are hidden assumptions behind it, and these assumptions could be wrong.

Let us first look at an example. The chemical compounds used to be classified into organic and inorganic, but there had never been a clear definition. The belief behind this definition is the vitalism, which assumes that there is a vital force, or "life force" in all the living organisms. Vitalism was proven to be wrong, with the synthesis of an organic compound urea from inorganic salts in 1828. A new definition of organic substance then becomes: a chemical substance that contains carbon atoms. However, CO, CO_2 contain carbon atoms, and the diamond contains nothing but carbon atoms, but these are not considered organic. The definition is then amended to: a chemical substance that contains carbon atoms and at least one hydrogen atom.

In my opinion, defining "organic" with an emphasis on carbon is *ad hoc*. The abundance of carbon atoms in organic substances is just a phenomenon, rather than the essence. The essence is the high complexity in structure due to the abundance in carbon. Because carbon has a valence of four, carbons are good building blocks to form a skeleton for large molecules. Large molecules are usually more complex than smaller ones. Although diamond has nothing but carbon, it is a crystal with a simple, periodic lattice structure. Because there is no clear boundary between the complex and the simple, there is no clear boundary between organic and inorganic substances.

The concept of intelligence is similar. There is a wrong assumption behind the concept—the human is fundamentally different from other things (living and nonliving together)—Humans are intelligent while rocks are not, plants are not, and even dogs and monkeys are not, simply because they are not humans. In my opinion, the essence of intelligence is also high complexity. Humans have high complexity in structure (especially in the brain), and in behavior as well. The complexity in behavior is the result of the complexity in structure. So intelligence is merely the description of systems with high complexity in behavior or functionality. Since there is no clear boundary between the complex and the simple, there is no clear boundary between intelligence and non-intelligence.

> *** Philosophical View**: What is intuition exactly?
>
> —My opinion: statistical sampling and interpolation
>
> Before the success of AlphaGo, it was believed that computers could not beat the best human Go players because humans use intuition in the play while computers cannot. The victory of AlphaGo was not due to merely brute-force calculations. Does it show that machines can possess intuition as well? For this we must first get clear what intuition is exactly. In fact, the success of AlphaGo indeed helps us to pin down what intuition really is. In my opinion, intuition is simply experience through statistical sampling. A beginner human Go player does not have much intuition about the board situations, while a professional player does, because the professional player has accumulated the statistical sampling unconsciously through years of experience. Now we know that statistical learning is not limited to humans. Machines may have "intuition" as well.

§2. The Tensor Data Structure

In recent years, using tensor as the basic data structure has become a new paradigm in machine learning and data science. In this context, a tensor is defined as a multi-dimensional array.

In computer science, many data structures are being used, like arrays, lists, stacks, trees and graphs. Multi-dimensional arrays naturally become the most general, versatile and convenient data structure to represent data in machine learning and data science. Vectors (as one dimensional arrays) and matrices (as two dimensional arrays) just become special cases of the general tensors.

2.1 AlphaGo

First let us look at how the input data are represented in AlphaGo. The Go board has a 19×19 grid. AlphaGo uses two 19×19 matrices to represent a board position, one for the black and the other for the white. Each matrix is binary, with 1 representing a stone present at this intersection and 0 representing absence of a stone of this color. This makes a tensor of dimensions $19 \times 19 \times 2$, with components ξ_{ijk}. If we fix k but vary i and

j, we have a 19×19 matrix, which we call it the kth plane. We also need to indicate whether it is black's turn or white's turn. To incorporate this information, we use another plane of a 19×19 matrix and duplicate the information. If it is black's turn, we fill all the entries with 1. Otherwise, fill the matrix with 0 (the actually implementation of AlphaGo is relative, 1 for "my turn" and 0 for "my opponent's turn"). This makes a tensor of $19 \times 19 \times 3$ to represent a board position. For the neural networks, a rolling history of moves is also needed. So AlphaGo keeps a history of 8 moves of each player to feed to the neural networks. Therefore, the inputs are tensors of dimensions $19 \times 19 \times 17$. The detailed implementation of AlphaGo is more complex. They decided to include additional features of each board position to simplify the computation. These feature planes include liberties, capture size, self-atari size, liberties after move, ladder capture, ladder escape, etc. (see Table 2.1). So the real feature dimension is 48. This is not absolutely necessary but only the implementation choice. The tensors fed to the neural networks have dimensions $19 \times 19 \times 48$. From the perspective of mathematics, this is a tensor product space of $V_1 \otimes V_2 \otimes V_3$, with $\dim V_1 = 19$, $\dim V_2 = 19$ and $\dim V_3 = 48$. In a later improved version, AlphaZero, they used input without additional features, namely dimensions $19 \times 19 \times 17$.

Table 2.1   Input features in AlphaGo

Feature	# of planes	Description
Stone color	3	Player stone / opponent stone / empty
Ones	1	A constant plane filled with 1
Turns since	8	How many turns since a move was played
Liberties	8	Number of liberties (empty adjacent points)
Capture size	8	How many opponent stones would be captured
Self-atari size	8	How many of own stones would be captured
Liberties after move	8	Number of liberties after this move is played
Ladder capture	8	Whether a move at this point is a successful ladder capture
Ladder escape	1	Whether a move at this point is a successful ladder escape
Sensibleness	1	Whether a move is legal and does not fill its own eyes
Zeros	1	A constant plane filled with 0
Play color	1	Whether current player is black

2.2   Images and Videos

Let us look at a particular example of an image. A gray-scale image of 1024×768 pixels can be represented by a 1024×768 matrix. A color image has three components of red, green and blue, which can be represented by a tensor ξ with dimensions $1024 \times 768 \times 3$, with components as ξ_{ijk} with $i \in \{1, \ldots, 1024\}, j \in \{1, \ldots, 768\}, k \in \{1, 2, 3\}$.

Suppose we have a video of 1800 frames in color, each frame being an image described as above. We can represent the video data with a tensor ζ of dimensions $1024 \times 768 \times 3 \times 1800$, with components ζ_{ijkt}, where t is the frame number, or discrete time.

In mathematics, ζ is viewed as an element in the tensor product space $V_1 \otimes V_2 \otimes V_3 \otimes V_4$, with $\dim V_1 = 1024$, $\dim V_2 = 768$, $\dim V_3 = 3$ and $\dim V_4 = 1800$.

2.3   Speech and Audio Applications

Next let us look at the data representation in speech processing applications, in particular, the speech to text conversion, in which text data are represented by a 1-dimensional tensor while the speech data are represented by a 2-dimensional tensor (spectrogram). Non-speech, general audio signals are just similar.

The text data are simpler. A text is a sequence of characters. With any character encoding, the text can be represented by an array of numbers, which can be viewed as a vector. The speech data in the analog form is a function of time $s(t)$ called the waveform. After sampling, it is converted to digital form, which is a function s_i of discrete time i, or a sequence of scalar values. In speech applications, it is more convenient to work in the frequency domain, rather in the time domain. That is, for a given sound, we analyze its frequency features, rather than the raw waveform. The waveform is converted to frequency features via discrete Fourier transform. A speech signal is a sequence of phonemes. A phoneme is the smallest unit of speech, like a consonant or a vowel. Different phonemes have different frequency features. Therefore, single Fourier transform of the entire audio clip does not make sense. What we do is to divide the audio into overlapping short-time frames. For each frame, we perform a short-time Fourier transform (STFT). The result is a frequency spectrum (the power intensity as a function of discrete frequency). After we have performed STFT for each of these frames for the entire audio, we obtain a sequence (indexed by discrete time) of spectra, which is called the spectrogram. The spectrogram for each frame is a vector, and the spectrogram for the entire audio is represented by a tensor (2-dimensional array, with one dimension in frequency and one dimension in time).

§3. TensorFlow and the Tensor Processing Unit (TPU)

TensorFlow is an open-source software library for machine learning developed by Google. TensorFlow 1.0 was released in 2017. It uses tensors as the basic data structure. It supports low level functions for tensor operations like addition, subtraction, tensor product, Hadamard product (element-wise multiplication) and tensor contraction. Most importantly, its high level functions facilitate easy implementation of deep neural networks. Because of the general data representation using tensors, TensorFlow is general purpose, rather than domain specific. It works for data from any field, like images, videos, speeches, online store user reviews, etc., when the data are represented using tensors.

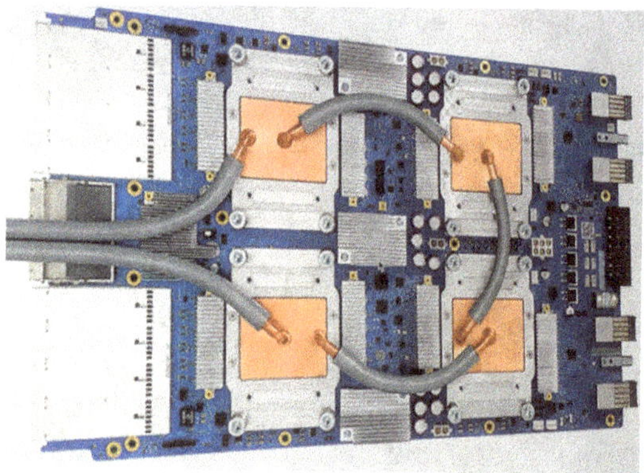

Figure 2.2   Google's Tensor Processing Unit 3.0

TensorFlow can run on multiple CPUs and GPUs. In 2016, Google announced its TPU (tensor processing unit), when the company said that the TPU had already been used inside their data centers for over a year. The TPU is an application-specific integrated circuit to accelerate tensor operations and is specifically designed for TensorFlow and machine learning applications. According to Norm Jouppi, AlphaGo was powered by TPUs in the match against the world champion Lee Sedol (note that the learning of AlphaGo was not real time during the match, but rather it was carried out "at home" in the days ahead of the match).

§4. Is Tensor in Machine Learning a Hype?

Someone has asked, "Multi-dimensional arrays are pretty simple, but is tensor just an unnecessary fancy name for it?" By calling them tensors, is it the same as calling water by the name *dihydrogen monoxide*? Another question is: are the tensors in machine learning the same as those in mathematics and physics?

Indeed there are situations that different concepts in different fields happen to get the same name by coincidence. The term "tensor" is also used in anatomy, to refer to a muscle that tightens or stretches a part of the body. Tensor in this sense has nothing to do with that in mathematics and physics. Is tensor in machine learning in a similar situation?

Sometimes in industry, names are manipulated as a marketing gimmick. The programming language JavaScript was officially called LiveScript when first shipped as part of a Navigator release in September 1995. The name was changed to JavaScript three months later. According to Wikipedia, "The choice of the JavaScript name has caused confusion, sometimes giving the impression that it is a spin-off of Java. Since Java was the hot new programming language at the time, this has been characterized as a marketing ploy by Netscape to give its own new language cachet."

Is tensor in machine learning a hype? Is it a marketing gimmick? The following is my opinion. The situation for the tensor in machine learning is different from tensor in anatomy, and also different from the situation of JavaScript vs. Java. Let us first look at the vector for example. As shown in Chapter 1, the vector can be defined abstractly via vector space as in Definition 11, but it can also be defined as a concrete example, as an n-tuple in Definition 10. A tensor can be defined abstractly via tensor product space as in Definition 5, but it can also be defined as a concrete example, namely multi-dimensional array as used in machine learning. Furthermore, when a basis for the abstract vector space is chosen, any finite dimensional vector can be represented by an n-tuple as components. When a basis for the abstract tensor space is chosen, any tensor can be represented by a multi-dimensional, multi-index array as components. When the basis changes, the components change according to those transformation laws.

The situation in machine learning is that we rarely have a need to change basis. So multi-dimensional arrays in machine learning are indeed tensors. This is not a marketing gimmick, but it is smart marketing.

Are the tensors in machine learning contravariant tensors or covariant tensors? Technically they are contravariant tensors. To see this, we use vectors for example again. The vectors defined as n-tuples are contravariant vectors. They have a natural basis $\mathbf{e}_1 = (1, 0, \ldots, 0), \mathbf{e}_2 = (0, 1, \ldots, 0), \ldots, \mathbf{e}_n = (0, 0, \ldots, 1)$, but we do have the liberty to change to a different basis any time. The tensors defined as multi-dimensional arrays have a natural basis too. We also have the liberty to change the basis and that will induce a coordinate change.

In the rest of this book, we are going to step away from this concrete definition in the context of machine learning, and explore more general, more abstract aspect of tensors in the context of mathematics and physics.

Chapter 3

Direct Sum Space $U \oplus V$

This short chapter belongs to linear algebra, rather than tensor algebra or multilinear algebra. It is a brief review. You can skip it if you are already familiar with it.

Given two vector spaces U and V over the same field F, we shall construct a new vector space $U \oplus V$, the direct sum space of U and V. We shall compare the differences between the direct sum space $U \oplus V$ and the tensor product space $U \otimes V$ in the future. The direct sum space $U \oplus V$ has a dimension of $\dim U + \dim V$ while the tensor product space $U \otimes V$ has a dimension of $\dim U \cdot \dim V$. We will find that the direct sum space is easy to construct, while the tensor product space is more difficult to construct.

In Chap. 7, we shall define the tensor algebra as the grand direct sum of tensor spaces of all types.

§1. The Elements

Given two vector spaces U and V over the same field F, we construct a new vector space $U \oplus V$, called the **direct sum space** of U and V. The set of elements of $U \oplus V$ is defined to be the Cartesian product $U \times V$, namely all the pairs $(\mathbf{u}, \mathbf{v})$ with $\mathbf{u} \in U$ and $\mathbf{v} \in V$.

§2. The Operations

Suppose U has addition $\mathfrak{d}$ and V has addition $\mathfrak{q}$ respectively. For $(\mathbf{u}_1, \mathbf{v}_1)$ and $(\mathbf{u}_2, \mathbf{v}_2)$, we define the addition:

$$(\mathbf{u}_1, \mathbf{v}_1) + (\mathbf{u}_2, \mathbf{v}_2) \overset{\text{def}}{=} (\mathbf{u}_1 \mathfrak{d} \mathbf{u}_2, \mathbf{v}_1 \mathfrak{q} \mathbf{v}_2),$$

and the scalar-vector multiplication:

$$a(\mathbf{u}_1, \mathbf{v}_1) \overset{\text{def}}{=} (a\mathbf{u}_1, a\mathbf{v}_1), \text{ for } a \in F.$$

It is easy to verify that $U \oplus V$ is a vector space under the addition and scalar-vector multiplication so defined.

§3. The Dimension of $U \oplus V$

All the elements $\{(\mathbf{u}, \mathbf{0}) | \mathbf{u} \in U\} \subseteq U \oplus V$ form a linear subspace of $U \oplus V$, where $\mathbf{0}$ is the zero vector of V. Similarly, all the elements $\{(\mathbf{0}, \mathbf{v}) | \mathbf{v} \in V\} \subseteq U \oplus V$ form a linear subspace of $U \oplus V$, where with some abuse of notation, $\mathbf{0}$ also denotes the zero element of U. We can identify the element $(\mathbf{u}, \mathbf{0})$ with the element $\mathbf{u} \in U$, and identify $(\mathbf{0}, \mathbf{v})$ with $\mathbf{v} \in V$. This way, U and V are embedded into $U \oplus V$ as linear subspaces. The following mappings are called **canonical injections**,

$$\iota_1 : U \to U \oplus V; \mathbf{u} \mapsto (\mathbf{u}, \mathbf{0}),$$
$$\iota_2 : V \to U \oplus V; \mathbf{v} \mapsto (\mathbf{0}, \mathbf{v}).$$

For any element $\mathbf{w} \in U \oplus V$, there are unique vectors $\mathbf{u} \in U$ and $\mathbf{v} \in V$ such that $\mathbf{w} = \mathbf{u} \oplus \mathbf{v}$. This is called the direct decomposition of $U \oplus V$ into U and V, with the projections called **canonical surjections**,

$$\pi_1 : U \oplus V \to U; (\mathbf{u}, \mathbf{v}) \mapsto \mathbf{u},$$
$$\pi_2 : U \oplus V \to V; (\mathbf{u}, \mathbf{v}) \mapsto \mathbf{v}.$$

Theorem. *Let $\{\mathbf{b}_1, \ldots, \mathbf{b}_m\}$ be a basis for U and $\{\mathbf{e}_1, \ldots, \mathbf{e}_n\}$ a basis for V. Then $\{(\mathbf{b}_1, \mathbf{0}), \ldots, (\mathbf{b}_m, \mathbf{0}), (\mathbf{0}, \mathbf{e}_1), \ldots, (\mathbf{0}, \mathbf{e}_n)\}$ is a basis for $U \oplus V$. Furthermore,*

$$\dim(U \oplus V) = \dim U + \dim V.$$

Remark. Let U, V be vector spaces over F. Consider the mapping $\oplus : U \times V \to U \oplus V; (\mathbf{u}, \mathbf{v}) \mapsto \mathbf{u} \oplus \mathbf{v} \stackrel{\text{def}}{=} (\mathbf{u}, \mathbf{v})$. This mapping $\oplus$ is not a bilinear mapping from U and V to $U \oplus V$.

The direct sum so defined is also called the external direct sum. For any vector space, the internal direct sum of two linear subspaces can also be defined. It can be proved that the external direct sum is equivalent (under isomorphism) to the internal direct sum.

Chapter 4

Gibbs Dyadics

In the previous chapter, given two vector spaces U and V, we have constructed the direct sum space $U \oplus V$ with the property $\dim(U \oplus V) = \dim U + \dim V$. For $\mathbf{u} \in U$ and $\mathbf{v} \in V$, we can have $\mathbf{u} \oplus \mathbf{v} \in U \oplus V$.

In this chapter, we would like to construct a space, which is in some sense a product of U and V. We call it the tensor product space of U and V and denote it by $U \otimes V$. It has the desired property $\dim(U \otimes V) = \dim U \cdot \dim V$. For $\mathbf{u} \in U$ and $\mathbf{v} \in V$, we also construct $\mathbf{u} \otimes \mathbf{v} \in U \otimes V$.

These are modern terms and notations, but the idea originated from Gibbs in the late 1800s. In Gibbs' own terms, $\mathbf{u} \otimes \mathbf{v}$ is called a dyad, denoted by $\mathbf{uv}$, just the juxtaposition of $\mathbf{u}$ and $\mathbf{v}$. The elements in $U \otimes V$ are called dyadics.

§1. What Is a Dyad?

Let V be a vector space over $\mathbb{R}$ and $\mathbf{u}, \mathbf{v} \in V$. A dyad $\mathbf{uv}$ is defined to be two vectors juxtaposed side by side.

Note a dyad $\mathbf{uv}$ can be defined in the general case for vectors $\mathbf{u} \in U$ and $\mathbf{v} \in V$, where U and V are different vector spaces, but in this chapter we just follow Gibbs and only treat the special case of $U = V$. The modern notation for $\mathbf{uv}$ is $\mathbf{u} \otimes \mathbf{v}$, which shall be called the tensor product. By using the juxtaposition $\mathbf{uv}$, we dodged the question "what is $\otimes$ in $\mathbf{u} \otimes \mathbf{v}$", which is harder to define. It is actually the tensor product and will be defined rigorously in the next chapter. We will use the juxtaposition notation throughout this chapter. We are guilty of hand-waving here, but we gain more intuition.

§2. When Are Two Dyads Equal?

We stipulate the following laws.

(1) If $\mathbf{u} = \mathbf{x}$ and $\mathbf{v} = \mathbf{y}$, then $\mathbf{uv} \overset{\text{def}}{=} \mathbf{xy}$.

(2) $(a\mathbf{u})\mathbf{v} \overset{\text{def}}{=} \mathbf{u}(a\mathbf{v})$, for any real number $a \in \mathbb{R}$.

For example, $(3\mathbf{u})\mathbf{v}$ and $\mathbf{u}(3\mathbf{v})$ are considered equal.

Remark. In general, $\mathbf{uv} \neq \mathbf{vu}$. We do not have to stipulate this as a rule, because this should be understood as long as we do not explicitly stipulate $\mathbf{uv} \overset{\text{def}}{=} \mathbf{vu}$, or prove $\mathbf{uv} = \mathbf{vu}$ based on other assumptions.

§3. What Are the Operations on Dyads?

We define the following operations on dyads.

(1) Scalar-dyad multiplication: namely, the multiplication of a scalar and a dyad is defined to be another dyad, for any $a \in \mathbb{R}$,

$$a(\mathbf{uv}) \overset{\text{def}}{=} (a\mathbf{u})\mathbf{v}.$$

For example, $3(\mathbf{uv})$ can be understood as either $(3\mathbf{u})\mathbf{v}$ or $\mathbf{u}(3\mathbf{v})$.

(2) The addition of two dyads $\mathbf{u}_1\mathbf{v}_1$ and $\mathbf{u}_2\mathbf{v}_2$ is

$$\mathbf{u}_1\mathbf{v}_1 + \mathbf{u}_2\mathbf{v}_2.$$

What type of object is this sum? Well, it is no longer a dyad (in general). It will be a new type of object called a dyadic, which has not been defined yet. We will define dyadics in the next section.

§4. What Is a Dyadic?

We just keep the form $\mathbf{u_1 v_1} + \mathbf{u_2 v_2}$ of two dyads with a "+" in between, and we call it a **dyadic**. A dyad is a special case of a dyadic, which is called a "**monomial**". So a dyadic can be a single dyad or a finite number of dyads connected by "+" signs, $\mathbf{u_1 v_1} + \ldots + \mathbf{u_k v_k}$. Do not ask what is the meaning of the "+" sign. It is just treated as a meaningless symbol and does not mean addition. The dyads connected by "+" as in $\mathbf{u_1 v_1} + \ldots + \mathbf{u_k v_k}$ are called the "formal sum" of these dyads. "Dyadic" is actually short for "**dyadic polynomial**", following Gibbs' terminology. The "formal sum" can be rigorously defined in modern language and be proved to be equivalent to the real addition. The real addition of two dyadics is defined in the next section.

§5. What Are the Operations on Dyadics?

We define the following operations on dyadics.

(1) The addition of two dyadics:
Given two dyadics $\mathbf{D}_1 = (\mathbf{u_1 v_1} + \ldots + \mathbf{u}_p \mathbf{v}_p)$ and $\mathbf{D}_2 = (\mathbf{x_1 y_1} + \ldots + \mathbf{x}_q \mathbf{y}_q)$, the addition (real sum) of them is defined to be the formal sum of them, namely, a dyadic made by the concatenation of the first dyadic, a "+" sign, and the second dyadic,

$$\mathbf{D}_1 + \mathbf{D}_2 \overset{\text{def}}{=} \mathbf{u_1 v_1} + \ldots + \mathbf{u}_p \mathbf{v}_p + \mathbf{x_1 y_1} + \ldots + \mathbf{x}_q \mathbf{y}_q.$$

Since the real sum (the addition operation) is defined as the formal sum, we do not have to distinguish them any more and this abuse of terms will not cause any confusion.

(2) Scalar-dyadic multiplication, namely, the multiplication of a scalar and a dyadic is defined to be another dyadic:

$$a(\mathbf{u_1 v_1} + \ldots + \mathbf{u}_k \mathbf{v}_k) \overset{\text{def}}{=} a(\mathbf{u_1 v_1}) + \ldots + a(\mathbf{u}_k \mathbf{v}_k),$$

for all $a \in \mathbb{R}$.

§6. When Are Two Dyadics Equal?

We stipulate the following laws.

(1) Commutativity of addition

Let $\mathbf{D}_1$ and $\mathbf{D}_2$ be two dyadics.

$$\mathbf{D}_1 + \mathbf{D}_2 = \mathbf{D}_2 + \mathbf{D}_1.$$

(2) Bilinearity of "juxtapose" operation (or distributive laws)

$$(a\mathbf{u}_1 + b\mathbf{u}_2)\mathbf{v} = a\mathbf{u}_1\mathbf{v} + b\mathbf{u}_2\mathbf{v},$$
$$\mathbf{u}(a\mathbf{v}_1 + b\mathbf{v}_2) = a\mathbf{u}\mathbf{v}_1 + b\mathbf{u}\mathbf{v}_2,$$

for all $a, b \in \mathbb{R}$.

Theorem 1. *All the dyadics over a vector space V with dyadic addition and scalar-dyadic multiplication form a vector space. We denote it by $V \otimes V$. Furthermore, if $\dim V = n$, then $\dim V \otimes V = n^2$.*

We show this only for the case of $\dim V = 3$. Let $\{\mathbf{e}_1, \mathbf{e}_2, \mathbf{e}_3\}$ be a basis for V. Then the vectors $\mathbf{u}$ and $\mathbf{v}$ can be written as

$$\mathbf{u} = u_1\mathbf{e}_1 + u_2\mathbf{e}_2 + u_3\mathbf{e}_3,$$
$$\mathbf{v} = v_1\mathbf{e}_1 + v_2\mathbf{e}_2 + v_3\mathbf{e}_3.$$

It is easy to show that

$$\begin{aligned}
\mathbf{u}\mathbf{v} = {} & u_1 v_1 \mathbf{e}_1\mathbf{e}_1 + u_1 v_2 \mathbf{e}_1\mathbf{e}_2 + u_1 v_3 \mathbf{e}_1\mathbf{e}_3 \\
& + u_2 v_1 \mathbf{e}_2\mathbf{e}_1 + u_2 v_2 \mathbf{e}_2\mathbf{e}_2 + u_2 v_3 \mathbf{e}_2\mathbf{e}_3 \\
& + u_3 v_1 \mathbf{e}_3\mathbf{e}_1 + u_3 v_2 \mathbf{e}_3\mathbf{e}_2 + u_3 v_3 \mathbf{e}_3\mathbf{e}_3.
\end{aligned}$$

Since any dyadic is the sum of dyads, it can be written as the sum of nine terms. The nine dyads $\{\mathbf{e}_1\mathbf{e}_1,\ \mathbf{e}_1\mathbf{e}_2,\ \mathbf{e}_1\mathbf{e}_3,\ \mathbf{e}_2\mathbf{e}_1,\ \mathbf{e}_2\mathbf{e}_2,\ \mathbf{e}_2\mathbf{e}_3,\ \mathbf{e}_3\mathbf{e}_1,\ \mathbf{e}_3\mathbf{e}_2,\ \mathbf{e}_3\mathbf{e}_3\}$ form a basis for this dyadic space as a vector space.

In modern language, the vector space $V \otimes V$ of all dyadics is called the tensor product space of V and itself (see Chap. 5).

§7. Matrix Representation

The first vector $\mathbf{u}$ can be represented by a column vector

$$\mathbf{u} = \begin{bmatrix} u_1 \\ u_2 \\ u_3 \end{bmatrix},$$

and the second vector $\mathbf{v}$ can be represented by a row vector

$$\mathbf{v} = \begin{bmatrix} v_1 & v_2 & v_3 \end{bmatrix}.$$

The dyad $\mathbf{uv}$ is then represented by the matrix multiplication of $\mathbf{u}$ and $\mathbf{v}$,

$$\mathbf{uv} = \begin{bmatrix} u_1 v_1 & u_1 v_2 & u_1 v_3 \\ u_2 v_1 & u_2 v_2 & u_2 v_3 \\ u_3 v_1 & u_3 v_2 & u_3 v_3 \end{bmatrix}.$$

The basis vectors can also be represented by matrices,

$$\mathbf{e_1 e_1} = \begin{bmatrix} 1 & 0 & 0 \\ 0 & 0 & 0 \\ 0 & 0 & 0 \end{bmatrix}, \quad \mathbf{e_1 e_2} = \begin{bmatrix} 0 & 1 & 0 \\ 0 & 0 & 0 \\ 0 & 0 & 0 \end{bmatrix}, \quad \text{etc.}$$

Since a dyadic is the sum of dyads, each dyadic can also be represented by a 3×3 matrix. The dyadic algebra is represented by the matrix algebra. The dyad $\mathbf{uv}$ is the matrix multiplication of $\mathbf{u}$ and $\mathbf{v}$.

Now the component definition of a tensor has found an explanation. The tensor is not defined as a matrix, but rather, *represented* by a matrix. The representation matrix of the tensor is relative to the basis of the dyadic space, which is a vector space. When the basis changes, the representation matrix changes accordingly, which we discuss in the following.

§8. Change of Coordinates

Let $\{\mathbf{e}_i\}$ be a basis for vector space V. Then $\{\mathbf{e}_i \mathbf{e}_j | i, j = 1, \ldots, n\}$ is a basis for the dyadic space $V \otimes V$. For dyad $\mathbf{uv}$, we have

$$\mathbf{uv} = \sum_{i,j=1}^{n} u^i v^j \mathbf{e}_i \mathbf{e}_j.$$

Suppose the basis $\{\mathbf{e}_i\}$ is changed to $\{\mathbf{e}_i'\}$,

$$\mathbf{e}_i' = \sum_{k=1}^{n} A_i^{\ k} \mathbf{e}_k,$$

$$\mathbf{e}_i = \sum_{k=1}^{n} \bar{A}_i^{\ k} \mathbf{e}_k',$$

where

$$\sum_{k=1}^{n} A_i^{\,k} \bar{A}_k^{\,j} = \delta_i^{\,j}.$$

The matrix $\bar{A}$ is the transpose of the inverse of A. The dyad $\mathbf{uv}$ can be represented under both the old basis or the new basis. That is,

$$\mathbf{uv} = \sum_{i,j}^{n} u^i v^j \mathbf{e}_i \mathbf{e}_j$$

$$= \sum_{i,j}^{n} \sum_{k,l}^{n} u^i v^j \bar{A}_i^{\,k} \bar{A}_j^{\,l} \mathbf{e}'_k \mathbf{e}'_l$$

$$= \sum_{k,l}^{n} (u')^k (v')^l \mathbf{e}'_k \mathbf{e}'_l.$$

Therefore,

$$(u')^k (v')^l = \sum_{i,j}^{n} u^i v^j \bar{A}_i^{\,k} \bar{A}_j^{\,l}.$$

This is the law of coordinate transformation of a dyad. Since any dyadic is the sum of dyads, and the dyadic is just a (contravariant) tensor (of degree 2), the law of coordinate transformation of a contravariant tensor of degree 2 is the same.

§9. What Are the Meanings of Dyadics?
Linear Transformations and Bilinear Forms

The dyadics have been defined as abstract forms (juxtapositions and formal sums) and it seems that we are only playing an abstract game. Do they have any connections to the real world? What do these dyadics represent?

In the very beginning, Gibbs did not invent them as an abstract, meaningless game. To him, dyads, as well as dyadics, are linear functions (in Gibbs' terminology, *vector functions*).

*** Review**: Linear Algebra—Linear functions and the dual space

Let V be a vector space over $\mathbb{R}$. A linear mapping $f : V \to \mathbb{R}$ is called a **linear function** (or **linear form**, or **covector**). All the linear functions form a vector space, which we call the **dual space** of V, denoted by V^*.

The dual space V^* is isomorphic to V and has the same dimension as V. The vectors in V^*, namely linear functions can be identified as vectors in V through the isomorphism.

Depending on the context, a vector can be viewed as a vector itself (contravariant vector), or can be viewed as a vector function (dual vector or covector). According to Gibbs, a dyadic can act on a single vector, or a pair of vectors, through dot product.

(1) Action of a dyadic on a single vector

We start with a dyad. Gibbs defines the dot product of a dyad $\mathbf{uv}$ and a vector $\mathbf{x}$ to be another vector,

$$(\mathbf{uv}) \cdot \mathbf{x} \overset{\text{def}}{=} \mathbf{u}\,(\mathbf{v} \cdot \mathbf{x}).$$

The result is a vector in the same (or opposite) direction as $\mathbf{u}$.

Therefore the meaning of the dyad $\mathbf{uv}$ is a linear transformation. As any dyadic is the sum of dyads, a dyadic is also a linear transformation.

Caution: not any linear transformation can be represented by a *dyad*, but any linear transformation can be represented by a *dyadic*, namely the sum of dyads.

(2) Action of a dyadic on a pair of vectors

Gibbs defines the dot product of a dyad $\mathbf{uv}$ and a pair of vectors, $\mathbf{x}$ and $\mathbf{y}$ to be a scalar.

$$\mathbf{y} \cdot (\mathbf{uv}) \cdot \mathbf{x} \overset{\text{def}}{=} (\mathbf{y} \cdot \mathbf{u})(\mathbf{v} \cdot \mathbf{x}).$$

Therefore the dyad here is interpreted as a bilinear form. As any dyadic is the sum of dyads, a dyadic is also a bilinear form.

Caution: not any bilinear form can be represented by a *dyad*, but any bilinear form can be represented by a *dyadic*, namely the sum of dyads.

*** Computer Science**: Partial application or "currying"

The above two views can be unified through "currying".

Any function, or mapping, can be viewed as a machine. It takes input and gives output. A bilinear form is a mapping $\Phi : V \times V \to \mathbb{R}$. We can write $\Phi(\cdot, \cdot)$, where the dots are place holders. This means this machine $\Phi(\cdot, \cdot)$ takes two vectors $(\mathbf{y}, \mathbf{x})$ as input and outputs a scalar $\Phi(\mathbf{y}, \mathbf{x})$. What if we feed $\Phi(\cdot, \cdot)$ with only one input vector $\mathbf{x}$? This is not the expected type of input according to the contract, but we will evaluate it anyway. After being fed with one input vector $\mathbf{x}$, the result is $\Phi(\cdot, \mathbf{x})$. We view $\Phi(\cdot, \mathbf{x})$ as a linear function, ready to output a scalar $\Phi(\mathbf{y}, \mathbf{x})$

upon a new input vector $\mathbf{y}$. This way, $\Phi(\cdot,\cdot)$ can be viewed as taking a vector $\mathbf{x}$ as input and giving a linear function $\Phi(\cdot,\mathbf{x})$ as output. If we identify V with its dual space V^*, we can view $\Phi(\cdot,\cdot)$ as taking a vector $\mathbf{x}$ as input and giving a vector $\Phi(\cdot,\mathbf{x})$ as output. So $\Phi(\cdot,\cdot)$ is viewed as a linear transformation.

This idea is used in computer science, functional programming in particular. When a function like $\Phi(\cdot,\cdot)$ takes two inputs by definition, we may supply with fewer inputs (e.g., just one input $\mathbf{x}$), and the result is another function $\Phi(\cdot,\mathbf{x})$ expecting the next input. This method is called partial application of a function, or "currying". The term is named for the logician Haskell Curry and has nothing to do with cooking.

§10. What Is the Nature of Dyadic Juxtaposition?

We have confessed that we committed a sin of hand-waving by defining a dyad as the *juxtaposition* of two vectors in Sec. 1. What is the nature of this juxtaposition exactly?

We have seen many different forms of multiplication. For example, the ordinary multiplication of two real numbers (yielding a real number), the multiplication of a scalar and a vector (yielding a vector), the dot product of two vectors (yielding a scalar), the cross product of two vectors (yielding a vector). What do they have in common? What is the essence of multiplication or product? We use $*$ to represent any type of multiplication. The essence is that a multiplication is a bilinear mapping

$$(*) : X \times Y \to Z.$$

By bilinear, we mean

$$(ax_1 + bx_2) * y = ax_1 * y + bx_2 * y,$$
$$x * (ay_1 + by_2) = ax * y_1 + bx * y_2,$$

for all $a, b \in \mathbb{R}$ and $x_1, x_2 \in X, y_1, y_2 \in Y$. This applies to all the examples of real number multiplication, multiplication of a scalar and a vector, the dot product of two vectors and the cross product of two vectors.

We have stipulated the laws defining when two dyads are equal and when two dyadics are equal. These laws stipulate that the dyadic juxtaposition has the properties of a bilinear mapping. It is some product in nature. This is the tensor product $\otimes$, which we shall define rigorously in the next chapter.

Chapter 5

Tensor Spaces (Tensor Product $U \otimes V$)

In this chapter, we shall present the rigorous and abstract definitions of tensor. This will be an axiomatic approach. Namely, we define some tensor space abstractly and the elements of this tensor space will be called tensors. This tensor space is related to two underlying vector spaces U and V and is called the tensor product space of U and V, denoted by $U \otimes V$. This definition of tensor space is in a similar position to Peano's axiomatic definition of vector space, which is Definition 11 in Chap. 1. Some familiar tensors will become examples or models of the tensor space in this definition.

The tensor product spaces of V and itself, such as $V \otimes V$, $V \otimes V^*$ and $V^* \otimes V^*$, namely tensor power spaces, will be discussed in the next chapter.

§1. Bilinear Mappings

Intuitively, a bilinear mapping is a mapping which is linear with respect to each variable.

Definition 1. (Bilinear mapping) Let U, V and W be vector spaces over the same field F. A mapping $\Phi : U \times V \to W$ is called a **bilinear mapping** if it is linear in each variable separately. Namely, for all $\mathbf{u}$, $\mathbf{u}_1$, $\mathbf{u}_2 \in U$, $\mathbf{v}$, $\mathbf{v}_1$, $\mathbf{v}_2 \in V$ and $a, b \in F$,

$$\Phi(a\mathbf{u}_1 + b\mathbf{u}_2, \mathbf{v}) = a\Phi(\mathbf{u}_1, \mathbf{v}) + b\Phi(\mathbf{u}_2, \mathbf{v}),$$
$$\Phi(\mathbf{u}, a\mathbf{v}_1 + b\mathbf{v}_2) = a\Phi(\mathbf{u}, \mathbf{v}_1) + b\Phi(\mathbf{u}, \mathbf{v}_2).$$

If $W = F$, a bilinear mapping $\Phi : U \times V \to F$ is called a **bilinear function**, or **bilinear form**.

Example 1. (Inner product) Let $\mathbf{u} = (x_1, \ldots, x_n)$, $\mathbf{v} = (y_1, \ldots, y_n) \in \mathbb{R}^n$. The inner product $\varphi : \mathbb{R}^n \times \mathbb{R}^n \to \mathbb{R}$, which is defined by

$$\varphi(\mathbf{u}, \mathbf{v}) \stackrel{\text{def}}{=} \langle \mathbf{u}, \mathbf{v} \rangle \stackrel{\text{def}}{=} x_1 y_1 + \ldots + x_n y_n,$$

is a bilinear form.

Example 2. (Cross product) Let $\mathbf{u} = (x_1, x_2, x_3)$, $\mathbf{v} = (y_1, y_2, y_3) \in \mathbb{R}^3$. The cross product $(\times) : \mathbb{R}^3 \times \mathbb{R}^3 \to \mathbb{R}^3$, defined by

$$\mathbf{u} \times \mathbf{v} \stackrel{\text{def}}{=} (x_2 y_3 - x_3 y_2, \; x_3 y_1 - x_1 y_3, \; x_1 y_2 - x_2 y_1),$$

is a bilinear mapping.

Example 3. Let $\mathbf{u} = (x_1, x_2, x_3) \in \mathbb{R}^3$ and $\mathbf{v} = (y_1, y_2, y_3) \in \mathbb{R}^3$. We choose two constant vectors $\mathbf{a} = (1, 2, -1)$ and $\mathbf{b} = (1, -1, 1)$ in $\mathbb{R}^3$. Using the inner product in $\mathbb{R}^3$, we define a mapping $\varphi : \mathbb{R}^3 \times \mathbb{R}^3 \to \mathbb{R}$ by

$$\varphi(\mathbf{u}, \mathbf{v}) \stackrel{\text{def}}{=} \langle \mathbf{a}, \mathbf{u} \rangle \langle \mathbf{b}, \mathbf{v} \rangle = (x_1 + 2x_2 - x_3)(y_1 - y_2 + y_3).$$

The mapping φ is a bilinear mapping.

Example 4. (Matrix product) Let $M_{m,k}$ be the vector space of all $m \times k$ matrices and $M_{k,n}$ be the vector space of all $k \times n$ matrices. The matrix multiplication $\varphi : M_{m,k} \times M_{k,n} \to M_{m,n}$ is a bilinear mapping.

Example 5. (Product of complex numbers) If the complex numbers $\mathbb{C}$ are viewed as a 2-dimensional vector space $\mathbb{R}^2$ over $\mathbb{R}$, then the complex number multiplication $() : \mathbb{R}^2 \times \mathbb{R}^2 \to \mathbb{R}^2$, $(x, y) \mapsto xy$ is a bilinear mapping. If $x = x_1 + ix_2$ and $y = y_1 + iy_2$, then

$$xy = (x_1 y_1 - x_2 y_2) + i(x_1 y_2 + x_2 y_1).$$

Example 6. (Product of column vectors and row vectors) Let $\mathbf{u} = (x_1, \ldots, x_m) \in \mathbb{R}^m$ and $\mathbf{v} = (y_1, \ldots, y_n) \in \mathbb{R}^n$. We can view $\mathbf{u}$ and $\mathbf{v}$ as column vectors. Namely,

$$\mathbf{u} = \begin{bmatrix} x_1 \\ \vdots \\ x_m \end{bmatrix}, \ \mathbf{v} = \begin{bmatrix} y_1 \\ \vdots \\ y_n \end{bmatrix}$$

are $m \times 1$ and $n \times 1$ matrices respectively.

We define $\varphi : \mathbb{R}^m \times \mathbb{R}^n \to M_{m,n}$, $\varphi(\mathbf{u}, \mathbf{v}) \overset{\text{def}}{=} \mathbf{u}\mathbf{v}^t$, where $\mathbf{v}^t$ is the transpose of $\mathbf{v}$, which is a row vector, or $1 \times n$ matrix. This is a special case of matrix multiplication in Example 4. The result

$$\mathbf{u}\mathbf{v}^t = \begin{bmatrix} x_1 y_1 & \cdots & x_1 y_n \\ \vdots & \vdots & \vdots \\ x_m y_1 & \cdots & x_m y_n \end{bmatrix}$$

is an $m \times n$ matrix with entries $A_{ij} = x_i y_j$. This mapping φ is a bilinear mapping.

Note that $\mathbf{u}\mathbf{v}^t$ and $\mathbf{u}^t\mathbf{v}$ are different. $\mathbf{u}\mathbf{v}^t$ is an $m \times n$ matrix, but $\mathbf{u}^t$ and $\mathbf{v}$ cannot be multiplied if $m \neq n$. $\mathbf{u}^t\mathbf{v}$ is a scalar if $m = n$.

Example 7. (Kronecker product of two matrices) Let

$$A = \begin{bmatrix} a_{11} & a_{12} \\ a_{21} & a_{22} \end{bmatrix} \in M_{2,2}$$

be a 2×2 matrix and

$$B = \begin{bmatrix} b_{11} & b_{12} & b_{13} \\ b_{21} & b_{22} & b_{23} \end{bmatrix} \in M_{2,3}$$

be a 2×3 matrix. The Kronecker product $\otimes : M_{2,2} \times M_{2,3} \to M_{4,6}$, $(A, B) \mapsto A \otimes B$ is defined as

$$A \otimes B \overset{\text{def}}{=} \begin{bmatrix} a_{11}B & a_{12}B \\ a_{21}B & a_{22}B \end{bmatrix} \overset{\text{def}}{=} \begin{bmatrix} a_{11}b_{11} & a_{11}b_{12} & a_{11}b_{13} & a_{12}b_{11} & a_{12}b_{12} & a_{12}b_{13} \\ a_{11}b_{21} & a_{11}b_{22} & a_{11}b_{23} & a_{12}b_{21} & a_{12}b_{22} & a_{12}b_{23} \\ a_{21}b_{11} & a_{21}b_{12} & a_{21}b_{13} & a_{22}b_{11} & a_{22}b_{12} & a_{22}b_{13} \\ a_{21}b_{21} & a_{21}b_{22} & a_{21}b_{23} & a_{22}b_{21} & a_{22}b_{22} & a_{22}b_{23} \end{bmatrix}.$$

This is a bilinear mapping. This can be generalized to the Kronecker product of two matrices of any size. If A is an $m \times n$ matrix and B is a $p \times q$ matrix, then $A \otimes B$ is an $mp \times nq$ matrix.

§2. Differences: Bilinear Mapping vs. Linear Mapping

> *** Review**: Linear Algebra—Definition of linear mapping
>
> Let V and W be vector spaces over the same field F. A mapping $\varphi : V \to W$ is called a **linear mapping**, if it satisfies the following conditions. For all $\mathbf{v}_1, \mathbf{v}_2, \mathbf{v} \in V$ and $a \in F$,
>
> $$\varphi(\mathbf{v}_1 + \mathbf{v}_2) = \varphi(\mathbf{v}_1) + \varphi(\mathbf{v}_2),$$
> $$\varphi(a\mathbf{v}) = a\varphi(\mathbf{v}).$$

Because a bilinear mapping is a mapping which is linear with respect to each variable, people tend to think that bilinear mappings are just similar to linear mappings. Yes, they are similar. However, significant differences between the two are often overlooked.

We have seen that number multiplication is a bilinear mapping. In general, a bilinear mapping is very much like multiplication. In the examples of bilinear mappings, we have seen inner product, cross product, matrix product and Kronecker product. These are all some sort of multiplications.

Let U, V and W be vector spaces over $\mathbb{R}$. Let $\varphi : U \to W$ be a linear mapping and $\Phi : U \times V \to W$ be a bilinear mapping.

Difference 1. Let $\mathbf{u} \in U, \mathbf{v} \in V$ and $a \in \mathbb{R}$. For the linear mapping φ,

$$\varphi(a\mathbf{u}) = a\varphi(\mathbf{u}),$$

but for the bilinear mapping Φ,

$$a\Phi(\mathbf{u}, \mathbf{v}) \neq \Phi(a\mathbf{u}, a\mathbf{v}) = a^2\Phi(\mathbf{u}, \mathbf{v}),$$
$$a\Phi(\mathbf{u}, \mathbf{v}) = \Phi(a\mathbf{u}, \mathbf{v}) = \Phi(\mathbf{u}, a\mathbf{v}).$$

A symmetric bilinear mapping $\Phi : V \times V \to W$ induces a mapping $Q : V \to W; Q(\mathbf{v}) = \Phi(\mathbf{v}, \mathbf{v})$. Q is rather quadratic in nature, because $Q(a\mathbf{v}) = a^2 Q(\mathbf{v})$.

Difference 2. The image of the linear mapping $\mathrm{Im}\varphi \subseteq W$ is always a linear subspace of W, while the image of the bilinear mapping $\mathrm{Im}\Phi \subseteq W$ may or may not be a linear subspace of W. The image of these mappings are defined by $\mathrm{Im}\varphi \overset{\text{def}}{=} \{\varphi(\mathbf{u}) | \mathbf{u} \in U\}$ and $\mathrm{Im}\Phi \overset{\text{def}}{=} \{\Phi(\mathbf{u}, \mathbf{v}) | \mathbf{u} \in U, \mathbf{v} \in V\}$.

Let us look at one example. Let V be the space of all 2×1 column vectors and $\mathbf{u}, \mathbf{v} \in V$. The product $\mathbf{u} \otimes \mathbf{v}$ defined by matrix multiplication $\mathbf{u}\mathbf{v}^t$ is a bilinear mapping, where $\mathbf{v}^t$ is a row vector (the transpose of $\mathbf{v}$).

In the space $M_{2,2}$ of 2×2 matrices, if a matrix

$$A = \begin{bmatrix} a_{11} & a_{12} \\ a_{21} & a_{22} \end{bmatrix}$$

is in the form of $\mathbf{u}\mathbf{v}^t$ with $\mathbf{u}$ and $\mathbf{v}$ being column vectors, then A is in the image of $\otimes$. Suppose

$$\mathbf{u} = \begin{bmatrix} u_1 \\ u_2 \end{bmatrix}, \mathbf{v} = \begin{bmatrix} v_1 \\ v_2 \end{bmatrix}.$$

Then

$$A = \mathbf{u}\mathbf{v}^t = \begin{bmatrix} u_1 v_1 & u_1 v_2 \\ u_2 v_1 & u_2 v_2 \end{bmatrix}.$$

It is easy to check $\det A = 0$. This means $\det A = 0$ is the necessary condition for A to be in Im$\otimes$.

For instance, the matrix

$$A = \begin{bmatrix} 3 & 2 \\ 9 & 6 \end{bmatrix}.$$

has $\det A = 0$ and can be decomposed into $A = \mathbf{u}\mathbf{v}^t = \mathbf{u} \otimes \mathbf{v}$ with $\mathbf{u} = [1, 3]^t$ and $\mathbf{v} = [3, 2]^t$. (Caution: for an $n \times n$ matrix A with $n > 2$, $\det A = 0$ is a necessary condition but not a sufficient condition for A to be decomposable. The necessary and sufficient condition is rank$A = 1$. See Sec. 5.)

Now look at another example. Let $\mathbf{x} = [1, 0]^t, \mathbf{y} = [0, 1]^t$.

$$\mathbf{x} \otimes \mathbf{x} = \begin{bmatrix} 1 & 0 \\ 0 & 0 \end{bmatrix} \in \text{Im}\otimes,$$

$$\mathbf{y} \otimes \mathbf{y} = \begin{bmatrix} 0 & 0 \\ 0 & 1 \end{bmatrix} \in \text{Im}\otimes.$$

The sum of these two matrices is

$$\mathbf{x} \otimes \mathbf{x} + \mathbf{y} \otimes \mathbf{y} = \begin{bmatrix} 1 & 0 \\ 0 & 1 \end{bmatrix} \notin \text{Im}\otimes,$$

because

$$\det \begin{bmatrix} 1 & 0 \\ 0 & 1 \end{bmatrix} \neq 0.$$

Therefore, the image of $\otimes$ is not a linear subspace of $M_{2,2}$.

The image of $\otimes$ contains all 2×2 matrices A such that $\det A = 0$. Hence $\mathrm{Im}\otimes$ is a 3-dimensional curved hypersurface in a 4-dimensional vector space described by equation

$$a_{11}a_{22} - a_{12}a_{21} = 0.$$

It is hard to visualize such a 3-dimensional curved hypersurface but if we fix one variable, say $a_{21} = 1$, then the projection is $a_{12} = a_{11}a_{22}$. This can be represented by a curved surface $z = xy$, which is known as a saddle surface, or a hyperbolic paraboloid in Figure 5.1.

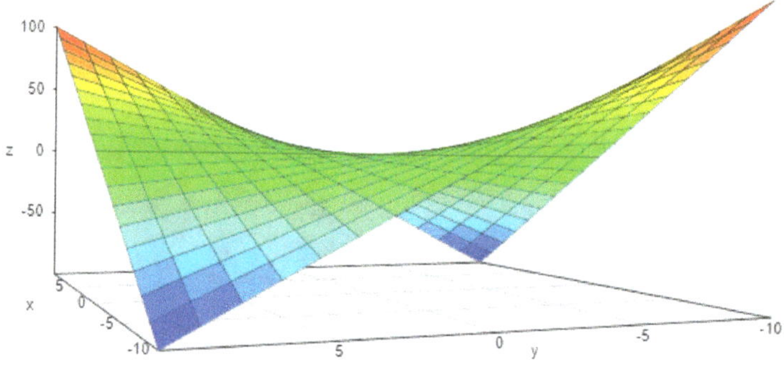

Figure 5.1 $\mathrm{Im}\otimes$ is a 3-dimensional hypersurface, but not a linear subspace. Its projection is a saddle surface.

> *** Review.** Linear Algebra—Definition of span
>
> Let S be a subset of vector space W. All linear combinations of the elements in S form a linear subspace of W, which is called the **span**, or **linear span**, or **linear closure** of S (or the **subspace spanned by** S, the **subspace generated by** S), and is denoted by $\langle S \rangle$ or $\mathrm{Span}(S)$.

For a bilinear mapping Φ, the image $\mathrm{Im}\Phi$ is not necessarily a linear subspace of W, but it can generate a linear subspace of W, denoted by $\langle \mathrm{Im}\Phi \rangle$. For a linear mapping $\varphi : U \to W$, it is always the case $\langle \mathrm{Im}\varphi \rangle = \mathrm{Im}\varphi$, which is a linear subspace of W.

Question. Let U, V and W be vector spaces over $\mathbb{R}$. Let $\varphi : U \to W$ be a linear mapping, and $\Phi : U \times V \to W$ a bilinear mapping. What are the maximum possible dimensions of $\mathrm{Im}\varphi$ and $\langle \mathrm{Im}\Phi \rangle$?

Answer. The maximum possible dimension of $\operatorname{Im}\varphi$ is $\dim U$. The maximum possible dimension of $\langle \operatorname{Im}\Phi \rangle$ is $\dim U \cdot \dim V$. In the special case of $U = V$, the maximum possible dimension of $\langle \operatorname{Im}\Phi \rangle$ is $(\dim U)^2$.

To prove these claims, we need the following theorems.

Theorem 1. (Linear extension of a linear mapping) *Let* $\{\mathbf{e}_1, \ldots, \mathbf{e}_n\}$ *be a basis of vector space* U. *Let* $\mathbf{w}_1, \ldots, \mathbf{w}_n \in W$ *be arbitrary vectors in* W. *There exists a unique linear mapping* $\varphi : U \to W$ *such that* $\varphi(\mathbf{e}_i) = \mathbf{w}_i, i = 1, \ldots, n$.

This is to say, to define a linear mapping $\varphi : U \to W$, it suffices to assign values to all the basis vectors $\varphi(\mathbf{e}_i) = \mathbf{w}_i, i = 1, \ldots, n$. The value for any other vectors can be linearly extended by

$$\varphi\left(\sum_i \lambda_i \mathbf{e}_i\right) = \sum_i \lambda_i \varphi(\mathbf{e}_i) = \sum_i \lambda_i \mathbf{w}_i.$$

If all $\mathbf{w}_1, \ldots, \mathbf{w}_n \in W$ are linearly independent, then $\operatorname{Im}\varphi$ has the maximum possible dimension $n = \dim U$.

Theorem 2. (Bilinear extension of a bilinear mapping) *Let* $\{\mathbf{b}_1, \ldots, \mathbf{b}_m\}$ *be a basis for* U, $\{\mathbf{e}_1, \ldots, \mathbf{e}_n\}$ *a basis for* V, *and* $\mathbf{w}_{ij} \in W, i = 1, \ldots, m; j = 1, \ldots, n$, *be arbitrary vectors in* W. *There exists a unique bilinear mapping* $\Phi : U \times V \to W$ *such that*

$$\Phi(\mathbf{b}_i, \mathbf{e}_j) = \mathbf{w}_{ij}, i = 1, \ldots, m; j = 1, \ldots, n.$$

This is to say, to define a bilinear mapping $\Phi : U \times V \to W$, it suffices to assign values to all the basis vector pairs $\Phi(\mathbf{b}_i, \mathbf{e}_j) = \mathbf{w}_{ij}$. The value for any other vector pairs can be bilinearly extended by

$$\Phi\left(\sum_i \lambda_i \mathbf{b}_i, \sum_j \mu_j \mathbf{e}_j\right) = \sum_{i,j} \lambda_i \mu_j \Phi(\mathbf{b}_i, \mathbf{e}_j) = \sum_{i,j} \lambda_i \mu_j \mathbf{w}_{i,j}.$$

Theorem 3. *Let U, V and W be finite dimensional vector spaces and $\Phi : U \times V \to W$ be a bilinear mapping. Then*

$$\dim \langle \mathrm{Im}\Phi \rangle \leq \dim U \cdot \dim V.$$

The equality holds if and only if the set of vectors $\{\Phi(\mathbf{b}_i, \mathbf{e}_j) \in W | i = 1, \ldots, m, \ j = 1, \ldots, n\}$ are linearly independent, where $\{\mathbf{b}_i | i = 1, \ldots, m\}$ is a basis for U and $\{\mathbf{e}_j | j = 1, \ldots, n\}$ is a basis for V.

If all $\mathbf{w}_{ij} \in W, i = 1, \ldots, m; j = 1, \ldots, n$, are linearly independent, then the dimension of $\langle \mathrm{Im}\Phi \rangle$ is $\dim U \cdot \dim V$.

Difference 3. Let $\varphi : U \to W$ be a linear mapping and $\Phi : U \times V \to W$ be a bilinear mapping. The maximum possible dimension of $\mathrm{Im}\varphi$ is $\dim U$. The maximum possible dimension of $\langle \mathrm{Im}\Phi \rangle$ is $\dim U \cdot \dim V$, which is much larger than the dimension of $U \oplus V$ in general, which is $\dim U + \dim V$.

§3. Multilinear Mappings

Bilinear mappings can be easily generalized to multilinear mappings.

Definition 2. (Multilinear mapping) Let $V_1, \ldots, V_p$ and W be vector spaces over the same field F. A mapping $\Phi : V_1 \times \cdots \times V_p \to W$ is called a **multilinear mapping** (or **p-linear mapping**) if it is linear in each variable separately. If $W = F$, a multilinear mapping $\Phi : V_1 \times \cdots \times V_p \to F$ is called a **multilinear function** (or **multilinear form**, or **p-linear form**).

§4. Tensor Product Space of Two Vector Spaces

Definition 3. (Tensor product space) Let U, V and W be vector spaces, and $\otimes : U \times V \to W$ be a bilinear mapping. The pair $(W, \otimes)$ is called a **tensor product space** (or simply **tensor space**) over the underlying vector spaces U and V, if they satisfy the following conditions:

(1) Generating property

$$W = \langle \mathrm{Im} \otimes \rangle \, ;$$

(2) Maximal span property

$$\dim W = \dim U \cdot \dim V.$$

The vectors in W are called **tensors** over U and V. The mapping $\otimes$ is called the **tensor multiplication** of two vectors, or **tensor product mapping**, or simply **tensor product**, or **tensor mapping**. W is often denoted by $U \otimes V$.

*** Methodology**: Constructive definition vs. axiomatic definition

We have defined the direct sum space $U \oplus V$ of two vector spaces U and V in Chap. 3. That is an example of definition by construction. Now we have a desire to define another vector space out of two given vector spaces U and V. We would call the new vector space the tensor product space $U \otimes V$.

An intuitive idea would be to construct such a space $U \otimes V$, which should have the property $\dim(U \otimes V) = \dim U \cdot \dim V$. In fact, we have done so in Chap. 4. That is the Gibbs dyadics. We used the "formal sum" in the construction, which is guilty of hand-waving. It is possible to define this rigorously, but it uses more abstract language (see the box at the end of this section). It is more difficult and more complex than the construction of the direct sum space.

Here we would rather go by an axiomatic definition. Rather than constructing such a tensor product space W starting from scratch, we would assume such a space W already exists. We only need to stipulate the characterizing properties, which are viewed as axioms. We can use an analogy to compare these two approaches—constructive vs. axiomatic. The constructive approach is like a bird building its nest from scratch using twigs. The axiomatic approach is like a hermit crab building its home by finding a scavenged shell of other species.

Even the axiomatic definition of tensor product space has quite many different but equivalent variations. We start with an easy one, and then compare it with others.

Definition 3 can be put in a slightly different but equivalent form in the following.

Definition 4. (Equivalent definition of tensor product space) Let $\otimes :$ $U \times V \to W$ be a bilinear mapping. The pair $(W, \otimes)$ is called a **tensor product space** of U and V, if it satisfies the following conditions:

(1′) Generating property

$$\langle \mathrm{Im}\otimes \rangle = W;$$

(2′) Maximal span property

$$\dim \langle \mathrm{Im}\otimes \rangle = \dim U \cdot \dim V.$$

Remark. The equivalence of the two definitions is trivial, because (1) and (1′) are the same, and (2′) is the trivial substitution of $\langle \mathrm{Im}\otimes \rangle$ for W in (2). Definition 1 focuses more on the space W (with W appearing on the left-hand-side of both equations), while Definition 2 focuses more on characterizing the bilinear mapping $\otimes$ (with $\otimes$ appearing on the left-hand-side of both equations).

Remark. Why is condition (1) needed at all? Is condition (2) alone, or condition (2′) alone be good enough for the definition? We have known, for any bilinear mapping $\otimes : U \times V \to W$,

$$\dim \langle \mathrm{Im}\otimes \rangle \leq \dim U \cdot \dim V.$$

At the heart of the definition of tensor product space is the "maximal span property". That is, the "tensor product mapping" can be spanned to the maximal dimension $\dim U \cdot \dim V$. This maximal span property should be literally

$$\dim \langle \mathrm{Im}\otimes \rangle = \dim U \cdot \dim V.$$

Condition (2) is not equivalent to condition (2′) by themselves. They are equivalent only under the premise of (1). Condition (1′) implies $\dim W \leq \dim U \cdot \dim V$, while condition (2′) implies $\dim W \geq \dim U \cdot \dim V$. Using the analogy of hermit crab mentioned above, the space W is just the scavenged shell that the hermit crab finds for its home. Condition (1′) says that scavenged shell W cannot be too big, while condition (2′) says that W cannot be too small. Together they imply $\dim W = \dim \langle \mathrm{Im}\otimes \rangle = \dim U \cdot \dim V$.

If condition (1) fails, it means that Im$\otimes$ does not generate W. Then there may exist a counterexample in which dim$W = $ dim$U \cdot$ dim$V$, but dim $\langle$Im$\otimes\rangle <$ dim$U \cdot$ dim$V$. This means a case in which W is the right size of dim$U \cdot$ dim$V$ while the mapping $\otimes$ is not spanned to the maximal possible dimension. Let us give a counterexample to illustrate this. Let $U = V = \mathbb{R}^3$ and $W = \mathbb{R}^9$. Condition (2) is clearly satisfied. We define the bilinear mapping $\otimes : U \times V \to W$ such that for any $\mathbf{u}, \mathbf{v} \in \mathbb{R}^3$, $\Phi(\mathbf{u}, \mathbf{v})$ is defined to be the cross product $\mathbf{u} \times \mathbf{v}$ into the first three dimensions of W, while keeping the coordinates of the other six dimensions as zero. In this case, Im$\otimes$ is a linear subspace of W. While dim$W = $ dim$U \cdot$ dim$V = 9$ holds, both condition (1′) and condition (2′) fail because dim $\langle$Im$\otimes\rangle = 3$.

Example 8. (Matrices) Let us revisit the bilinear mapping in Example 6, where $\mathbf{u} = (x_1, \ldots, x_m) \in \mathbb{R}^m$ and $\mathbf{v} = (y_1, \ldots, y_n) \in \mathbb{R}^n$. We view $\mathbf{u}$ and $\mathbf{v}$ as column vectors. Namely,

$$
\mathbf{u} = \begin{bmatrix} x_1 \\ \vdots \\ x_m \end{bmatrix}, \quad \mathbf{v} = \begin{bmatrix} y_1 \\ \vdots \\ y_n \end{bmatrix}
$$

are $m \times 1$ matrix and $n \times 1$ matrix respectively. We define $\otimes : \mathbb{R}^m \times \mathbb{R}^n \to M_{m,n}$,

$$
\mathbf{u} \otimes \mathbf{v} = \mathbf{u}\mathbf{v}^t = \begin{bmatrix} x_1 y_1 & \cdots & x_1 y_n \\ \vdots & \vdots & \vdots \\ x_m y_1 & \cdots & x_m y_n \end{bmatrix},
$$

an $m \times n$ matrix with entries $A_{ij} = x_i y_j$. $(M_{m,n}, \otimes)$ is a tensor product space of $\mathbb{R}^m$ and $\mathbb{R}^n$.

Now we have an answer to the question: "Are tensors matrices?" The answer is yes, but we see that this is just one model of tensors. Furthermore, we have made the tensor product mapping explicit. This also offers an explanation why the matrices as tensors have to change coordinates when the bases in the vector spaces $\mathbb{R}^m$ and $\mathbb{R}^n$ are changed.

Example 9. (Gibbs dyadics) The Gibbs dyadics described in Chap. 4 readily become an example of tensor product. Given two vector spaces U and V, the tensor product space $U \otimes V$ is the vector space of all the dyadics in the form of $\mathbf{u}_1\mathbf{v}_1 + \mathbf{u}_2\mathbf{v}_2 + \ldots + \mathbf{u}_k\mathbf{v}_k$ with $\mathbf{u}_1, \ldots, \mathbf{u}_k \in U$, $\mathbf{v}_1, \ldots, \mathbf{v}_k \in V$. Namely,

$$
U \otimes V = \langle \{\mathbf{u}\mathbf{v} | \mathbf{u} \in U, \mathbf{v} \in V\} \rangle,
$$

where $\mathbf{uv}$ is a dyad. Note the notation $\langle S \rangle$ means the linear subspace spanned by the set S. The tensor product mapping $\otimes$ is defined by $(\mathbf{u}, \mathbf{v}) \mapsto \mathbf{u} \otimes \mathbf{v} \overset{\text{def}}{=} \mathbf{uv}$.

Example 10. (Bilinear forms) Let U and V be vector spaces over the same field F, and U^* and V^* be the dual spaces. Let $W = L(U, V; F)$ be the vector space of all bilinear forms $\xi : U \times V \to F$. We define a mapping $\otimes : U^* \times V^* \to W$; $(f, h) \mapsto f \otimes h \overset{\text{def}}{=} \xi$ such that for all $\mathbf{u} \in U$ and $\mathbf{v} \in V$,

$$(f \otimes h)(\mathbf{u}, \mathbf{v}) \overset{\text{def}}{=} \xi(\mathbf{u}, \mathbf{v}) \overset{\text{def}}{=} f(\mathbf{u})h(\mathbf{v}).$$

The vector space of all bilinear forms $L(U, V; F)$ together with mapping $\otimes$ is a tensor product of U^* and V^*.

$$W = L(U, V; F) = U^* \otimes V^* = \langle \{ f \otimes h \, | \, f \in U^*, h \in V^* \} \rangle.$$

This is a special case for Definition 4 in Chap. 1. Definition 4 is actually a model of tensors in the abstract sense.

Example 11. (Quadratic forms) Let V be a vector space over $\mathbb{R}$, and V^* be its dual space. Let Q be the vector space of all quadratic forms $\zeta : V \to \mathbb{R}$. We define a mapping $\otimes : V^* \times V^* \to Q$; $(f, h) \mapsto f \otimes h \overset{\text{def}}{=} \zeta$ such that for all $\mathbf{v} \in V$,

$$(f \otimes h)(\mathbf{v}) \overset{\text{def}}{=} \zeta(\mathbf{v}) \overset{\text{def}}{=} f(\mathbf{v})h(\mathbf{v}).$$

The vector space of all quadratic forms Q together with mapping $\otimes$ is a tensor product of V^* and V^*,

$$Q = V^* \otimes V^* = \langle \{ f \otimes h \, | \, f, h \in V^* \} \rangle.$$

Students in physics may have a question: what is the moment of inertia tensor exactly? How is it defined as a tensor, and why is it a tensor? Again, traditional physics textbooks call it a tensor just because it is introduced as a matrix. An answer in the modern context is that a moment of inertia tensor $\mathscr{I}$ can be viewed either as a quadratic form or a linear mapping. As a quadratic form, $\mathscr{I}$ maps the angular velocity vector $\boldsymbol{\omega}$ to a scalar (2 times kinetic energy). When $\mathscr{I}$ applies to a unit vector $\mathbf{n}$ in the direction of $\boldsymbol{\omega}$, it yields the scalar moment of inertia I about the axis $\mathbf{n}$. As a linear mapping, $\mathscr{I}$ maps the angular velocity vector $\boldsymbol{\omega}$ to the angular momentum vector $\mathbf{L}$ of a rigid body (see more in Chap. 8)

Example 12. (Linear mappings) Let U and V be vector spaces over the same field F, and U^* and V^* be the dual spaces. Let W be the vector space of all linear mappings $\varphi : V \to U$. Suppose $\mathbf{u} \in U, f \in V^*$. We define $\mathbf{u} \otimes f$ to be a linear mapping $\varphi \in W$ such that

$$(\mathbf{u} \otimes f)(\mathbf{x}) \overset{\text{def}}{=} \varphi(\mathbf{x}) \overset{\text{def}}{=} f(\mathbf{x})\mathbf{u} \text{ for all } \mathbf{x} \in V.$$

Note that $f \in V^*$ is a linear form and $f(\mathbf{x})$ is a scalar.

Any linear mapping in W is the linear combination of mappings in the form of $\mathbf{u} \otimes f$. $W = U \otimes V^*$ is the tensor product space of U and V^*.

$$W = U \otimes V^* = \langle \{ \mathbf{u} \otimes f \,|\, \mathbf{u} \in U, f \in V^* \} \rangle .$$

Example 13. (Meanings of Gibbs dyadics) For Gibbs dyadics $\sum \mathbf{u}_i \mathbf{v}_i$, if each $\mathbf{v}_i$ is identified with a dual vector in V^*, then the dyadic is a tensor in the space $U \otimes V^*$, namely all linear mappings $\varphi : V \to U$. If each $\mathbf{u}_i$ is identified with a dual vector in U^*, then the dyadic is a tensor in the space $V \otimes U^*$, namely all linear mappings $\psi : U \to V$. If each $\mathbf{u}_i$ is identified with a dual vector in U^*, and each $\mathbf{v}_i$ is identified with a dual vector in V^*, then the dyadic is a tensor in the space $U^* \otimes V^*$, namely all bilinear forms $\xi : U \times V \to F$.

Remark. Disambiguation: Subtle meanings of the "tensor product"

(1) It may mean the "tensor product mapping"

$$\otimes : U \times V \to W.$$

(2) It may mean the "value of this mapping"—tensor product of two vectors $\mathbf{u} \in U$ and $\mathbf{v} \in V$,

$$\mathbf{u} \otimes \mathbf{v} \in W,$$

which is a vector in W. Any vector in W is called a tensor, but not all tensors in W is the tensor product of two vectors in the form of $\mathbf{u} \otimes \mathbf{v}$. The tensors in the form of $\mathbf{u} \otimes \mathbf{v}$ are called decomposable tensors (dyads). In general, any tensor (dyadic) in W is the sum of decomposable tensors (dyads).

(3) It may mean the "tensor product space" $U \otimes V$. Note in this sense, $\otimes$ is not an operator to operate on two vector spaces. The vector spaces U and V are not operands. $U \otimes V$ is treated as a single symbol made of three characters. It is the same as a single character W. $U \otimes V$ is just for mnemonic purposes to remind us that there is a mapping $\otimes : U \times V \to W$.

(4) It may mean the "tensor product" of other objects, which yet to be defined later, e.g., the tensor product of two tensors. This is because a tensor space is also a vector space after all. We may also define the tensor product of two linear transformations, or the tensor product of two matrices (also called the Kronecker product).

Theorem 4. *Let U and V be vector spaces over the same field F. There exists a vector space W and a mapping $\otimes : U \times V \to W$ such that $(W, \otimes)$ is a tensor product space of U and V. Moreover, the tensor product $(W, \otimes)$ of U and V is unique up to isomorphism.*

One way to prove the existence of tensor product in Theorem 4 is by construction of a model. Basically it is the space of Gibbs dyadics made rigorous in the modern language using free vector space and quotient space. Because of the uniqueness of tensor product, some authors choose to use this model, or any other model as the definition of tensor product. Moreover, the notation $U \otimes V$ (rather than W) has been adopted in literature for "*the* tensor product space" of U and V, and $\otimes$ for "*the* tensor product mapping".

Theorem 5. *Let $(U \otimes V, \otimes)$ be the tensor product space of U and V. Suppose $\mathbf{b}_1, \ldots, \mathbf{b}_m \in U$ are linearly independent and $\mathbf{e}_1, \ldots, \mathbf{e}_n \in V$ are linearly independent. Then $\mathbf{b}_i \otimes \mathbf{e}_j$, $i = 1, \ldots m$, $j = 1, \ldots, n$ are linearly independent in $U \otimes V$.*

Corollary. *If $\{\mathbf{b}_1, \ldots, \mathbf{b}_m\}$ is a basis for U and $\{\mathbf{e}_1, \ldots, \mathbf{e}_n\}$ is a basis for V, then $\{\mathbf{b}_i \otimes \mathbf{e}_j \,|\, i = 1, \ldots, m, j = 1, \ldots, n\}$ is a basis for $U \otimes V$.*

Condition (2) in Definition 3 explicitly involves the dimensions of spaces, which is not considered elegant. Moreover, this definition does not apply to infinite dimensional vector spaces. The maximal span property can be expressed in an alternative way.

Definition 5. (Alternative definition of tensor product) Let U, V and W be vector spaces, and $\otimes : U \times V \to W$ a bilinear mapping. The pair $(W, \otimes)$ is called a tensor product space of U and V, if they satisfy the following conditions:

(1) Generating property:

$$W = \langle \mathrm{Im} \otimes \rangle \, ;$$

(2) Maximal span property:

If $\mathbf{b}_1, \ldots, \mathbf{b}_m \in U$ are linearly independent and $\mathbf{e}_1, \ldots, \mathbf{e}_n \in V$ are linearly independent, then $\mathbf{b}_i \otimes \mathbf{e}_j \in W$, $i = 1, \ldots m$, $j = 1, \ldots, n$ are linearly independent.

Theorem 6. *Let $(U \otimes V, \otimes)$ be the tensor product of vector spaces U and V. Then $x \otimes y = 0$ implies $x = 0$ or $y = 0$.*

The following theorem is another characterization of tensor product, which does not explicitly involve the basis or dimension. It is called the universal factoring property, or simply universal property.

Theorem 7. (Unique universal factoring property) *Let $(W, \otimes)$ be a tensor product space over U and V. For any vector space X and any bilinear mapping $\Psi : U \times V \to X$, there exists a unique linear mapping $\varphi : W \to X$ such that*

$$\Psi = \varphi \circ \otimes.$$

We denote $W = U \otimes V$.

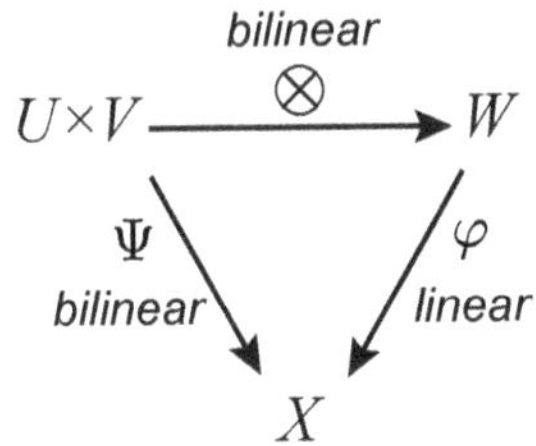

Figure 5.2   Universal property

The proof of this theorem can be found in most textbooks and hence is omitted. However, it is important to gain some insight about the essence of this theorem. Why does the tensor product mapping $\otimes$ guarantees the universal property? Now assume the tensor product properties of $\otimes$ are violated and let us see what happens.

(1) First assume that $\otimes : U \times V \to W$ does not have the maximum span property: $\dim W = p < \dim U \cdot \dim V = mn$, which is condition (2) in Definition 3. Then we can find a vector space X of dimension mn and a bilinear mapping $\Psi : U \times V \to X$ such that $\dim \langle \mathrm{Im}\Psi \rangle = mn$.

In such a case, there does not exist a linear mapping $\varphi : W \to X$ such that $\Psi = \varphi \circ \otimes$, because $\langle \mathrm{Im}\Psi \rangle$ has a higher dimension mn than $\dim W < mn$.

(2) Assume that $\otimes : U \times V \to W$ satisfies the maximum span property but does not satisfy the generating property $\langle \mathrm{Im}\otimes \rangle = W$. This means $\dim \langle \mathrm{Im}\otimes \rangle = mn$ but $\dim W > mn$. In such a case, there exists a linear mapping $\varphi : W \to X$ such that $\Psi = \varphi \circ \otimes$, but the linear mapping φ is not unique.

Therefore the universal property is equivalent to the maximal span property. The unique universal factoring property is equivalent to the conditions (1′) and (2′) in Definition 4.

The following lists some alternative definitions of tensor product in the boxes.

*** Equivalent Definition**: Tensor product—using universal factoring property

$(W, \otimes)$ is called a **tensor product space** of U and V, denoted by $W = U \otimes V$, if $\otimes : U \times V \to W$ is a bilinear mapping and satisfies the following conditions:

(1′) Generating property

$$\langle \mathrm{Im}\otimes \rangle = W;$$

(2′) Universal factoring property:

For any vector space X and any bilinear mapping $\Psi : U \times V \to X$, there exists a linear mapping $\varphi : W \to X$ such that

$$\Psi = \varphi \circ \otimes.$$

---

* **Equivalent Definition**: Tensor product—using unique universal factoring property

$(W, \otimes)$ is called a **tensor product space** of U and V, denoted by $W = U \otimes V$, if $\otimes : U \times V \to W$ is a bilinear mapping and satisfies the following condition:

For any vector space X and any bilinear mapping $\Psi : U \times V \to X$, there exists a *unique* linear mapping $\varphi : W \to X$ such that

$$\Psi = \varphi \circ \otimes.$$

* **Constructive Definition**: Tensor product

Let U and V be vector spaces over the same field F. Let $\mathscr{V}_F \langle U \times V \rangle$ be the free vector space generated by $U \times V$. Let Z be the subspace of $\mathscr{V}_F \langle U \times V \rangle$ generated by all the elements of the form

$$a(\mathbf{u}_1, \mathbf{v}) + b(\mathbf{u}_2, \mathbf{v}) - (a\mathbf{u}_1 + b\mathbf{u}_2, \mathbf{v}),$$
$$a(\mathbf{u}, \mathbf{v}_1) + b(\mathbf{u}, \mathbf{v}_2) - (\mathbf{u}, a\mathbf{v}_1 + b\mathbf{v}_2),$$

for all $a, b \in F$, $\mathbf{u}, \mathbf{u}_1, \mathbf{u}_2 \in U$ and $\mathbf{v}, \mathbf{v}_1, \mathbf{v}_2 \in V$.

The quotient space

$$U \otimes V = \frac{\mathscr{V}_F \langle U \times V \rangle}{Z}$$

is called the **tensor product** of U and V. The elements in $U \otimes V$ are called **tensors** over U and V.

Define a mapping $\otimes : U \times V \to U \otimes V$ such that for all $\mathbf{u} \in U$ and $\mathbf{v} \in V$, $(\mathbf{u}, \mathbf{v}) \mapsto \mathbf{u} \otimes \mathbf{v} \overset{\text{def}}{=} [(\mathbf{u}, \mathbf{v})]$, where $[(\mathbf{u}, \mathbf{v})]$ is the equivalence class of $(\mathbf{u}, \mathbf{v})$ in $\mathscr{V}_F \langle U \times V \rangle$ defined by the subspace Z. This mapping is a bilinear mapping and is called the **canonical bilinear mapping**.

See the *Encyclopedic Dictionary of Mathematics* [Mathematical Society of Japan (1993)]; see also [Bourbaki (1942); Roman (2005)].

See Appendix 1 for free vector spaces.

§5. Decomposable Tensors

Let U, V and W be vector spaces and $\Phi : U \times V \rightarrow W$ be a bilinear mapping. We have already emphasized in Sec. 2 that the image of the bilinear mapping $\mathrm{Im}\,\Phi \subseteq W$ may not be a linear subspace of W. This means that not every vector in W can be expressed in the form $\Phi(\mathbf{u}, \mathbf{v})$, for some $\mathbf{u} \in U$ and $\mathbf{v} \in V$. This leads to the definition of an important concept in the following.

Definition 6. (Decomposable tensor) Let $(U \otimes V, \otimes)$ be a tensor product of U and V. For a tensor $w \in U \otimes V$, if there exist $\mathbf{u} \in U$ and $\mathbf{v} \in V$ such that $w = \mathbf{u} \otimes \mathbf{v}$, then w is called a **decomposable tensor**.

Suppose $w \in V \otimes V$ is represented by matrix A. From the discussion in Sec. 2, we know that $\det A = 0$ is a necessary condition, but not a sufficient condition for $\dim V \geq 3$.

Theorem 8. *Suppose $w \in U \otimes V$ is represented by matrix A. w is a decomposable tensor if and only if*

$$\mathrm{rank}A = 1.$$

§6. Tensor Product of Linear Mappings

Definition 7. Let U_1, U_2, V_1, V_2 be vector spaces and $\varphi : U_1 \rightarrow V_1$, $\psi : U_2 \rightarrow V_2$ be linear mappings. There exists a unique linear mapping $\zeta : U_1 \otimes U_2 \rightarrow V_1 \otimes V_2$ such that for all $\mathbf{u}_1 \in U_1$ and $\mathbf{u}_2 \in U_2$,

$$\zeta(\mathbf{u}_1 \otimes \mathbf{u}_2) = \varphi(\mathbf{u}_1) \otimes \psi(\mathbf{u}_2).$$

ζ is called the **linear mapping induced** by φ and ψ, or the **tensor product** of φ and ψ.

§7. Tensor Product Space of Multiple Vector Spaces

Having defined the tensor product $V_1 \otimes V_2$ of two vector spaces, we can define the tensor product of $V_1 \otimes V_2$ and a third vector space V_3, namely $(V_1 \otimes V_2) \otimes V_3$. Alternatively, we can also define $V_1 \otimes (V_2 \otimes V_3)$. There is an isomorphism $(V_1 \otimes V_2) \otimes V_3 \to V_1 \otimes (V_2 \otimes V_3)$ that assigns $(v_1 \otimes v_2) \otimes v_3$ to $v_1 \otimes (v_2 \otimes v_3)$, $v_i \in V_i$ for $i = 1, 2, 3$.

Hence we can identify $(V_1 \otimes V_2) \otimes V_3$ with $V_1 \otimes (V_2 \otimes V_3)$ and simply denote them by $V_1 \otimes V_2 \otimes V_3$ without parentheses. This can be easily generalized to the tensor product of any number of vector spaces $V_1 \otimes \ldots \otimes V_m$.

Another approach is to define the tensor product of multiple vector spaces starting from scratch as follows, which is similar to the definition of the tensor product of two vector spaces, with the latter being a special case.

Definition 8. (Tensor product space of multiple vector spaces) Let $V_1, \ldots, V_m, W$ be vector spaces, and $\otimes : V_1 \times \ldots \times V_m \to W$ a multilinear mapping. The pair $(W, \otimes)$ is called a **tensor product space** of $V_1, \ldots, V_m$, if they satisfy the following conditions:
(1) Generating property:

$$W = \langle \mathrm{Im} \otimes \rangle ;$$

(2) Maximal span property:

$$\dim W = \prod_{i=1}^{m} \dim V_i.$$

W is often denoted by $V_1 \otimes \ldots \otimes V_m$.

All discussions about tensor product space $U \otimes V$ in Sec. 4 can be generalized to the tensor product space $V_1 \otimes \ldots \otimes V_m$.

§8. Vector-valued Tensors—The Most General Model

Let $U_1, \ldots, U_p, V_1, \ldots, V_q$ and X be vector spaces over the same field F. Consider the vector space W of all multilinear mappings

$$\Phi : V_1 \times \ldots \times V_q \to X.$$

Let $f_1 \in V_1^*, \ldots f_q \in V_q^*$ and $\mathbf{x} \in X$. We define

$$\otimes : X \times V_1^* \times \ldots \times V_q^* \to W;$$

$$\mathbf{x} \otimes f_1 \otimes \ldots \otimes f_q \overset{\text{def}}{=} \mathbf{x} f_1(\cdot) \ldots f_q(\cdot). \tag{5.1}$$

Hence $(W, \otimes)$ is a model of the tensor product space

$$W = X \otimes V_1^* \otimes \ldots \otimes V_q^*.$$

This means, the tensor product space $W = X \otimes V_1^* \otimes \ldots \otimes V_q^*$ is generated by all the tensors in Eq. 5.1.

If U_i is viewed as the dual space of U_i^*, then a multilinear mapping

$$\Psi : U_1^* \times \ldots \times U_p^* \to X$$

is a model of a tensor in the tensor product space

$$W = X \otimes U_1 \otimes \ldots \otimes U_p.$$

Most generally, a multilinear mapping

$$\Pi : U_1^* \times \ldots \times U_p^* \times V_1 \times \ldots \times V_q \to X \tag{5.2}$$

is a model of a tensor in the tensor product space

$$W = X \otimes U_1 \otimes \ldots \otimes U_p \otimes V_1^* \otimes \ldots \otimes V_q^*.$$

This answers one of the questions in Chap. 1. When we discussed Definition 4: a tensor is a multilinear form

$$\Phi : V^* \times \ldots \times V^* \times V \times \ldots \times V \to \mathbb{R},$$

we had a question: why does the codomain have to be the real numbers $\mathbb{R}$? The answer is that it does not have to. It can be any vector space X. That Definition 4 is only one model of tensors. In general, a tensor can be any general vector-valued multilinear mapping as in Eq. 5.2, or as a special case,

$$\Phi : V^* \times \ldots \times V^* \times V \times \ldots \times V \to V.$$

As a special case, when $p = 0$, $q = 1$, a tensor

$$\Pi : V_1 \to X$$

is a linear mapping from V_1 to X, which is an element in the tensor product space $X \otimes V_1^*$.

As another special case, when $p = 0$, $q = 2$ and $X = \mathbb{R}$, a tensor

$$\Pi : V_1 \times V_2 \to \mathbb{R}$$

is a bilinear form, which is an element in the tensor product space $V_1^* \otimes V_2^*$.

Chapter 6

Tensor Spaces (Tensor Power $V^{\otimes(p,q)}$)

A special type of tensor product is important in many applications. That is the tensor product of multiple copies of a vector space V itself (including its dual space V^*). They are called the tensor power space (or simply tensor power, or tensor space) over V. For example, $V \otimes V$, $V \otimes V \otimes V^*$, etc.

§1. Tensor Spaces (Tensor Power Spaces)

Definition 1. (Contravariant, covariant and mixed tensor spaces) The tensor product spaces

$$V^{\otimes(2,0)} \stackrel{\text{def}}{=} V^{\otimes 2} \stackrel{\text{def}}{=} V \otimes V,$$

$$V^{\otimes(0,2)} \stackrel{\text{def}}{=} (V^*)^{\otimes 2} \stackrel{\text{def}}{=} V^* \otimes V^*,$$

$$V^{\otimes(1,1)} \stackrel{\text{def}}{=} V \otimes V^*,$$

are called **contravariant tensor space**, **covariant tensor space**, and **mixed tensor space** (of degree 2) over V respectively. Their elements are called **contravariant tensors**, **covariant tensors** and **mixed tensors** (of degree 2). Alternatively, these tensor spaces are also denoted by $T_0^2(V), T_1^1(V), T_2^0(V)$ respectively.

Vectors in V are also called **contravariant vectors**, while vectors in V^* are called **covariant vectors**, or **covectors**. These terms have to do with the coordinate transformation of these vectors. When the old basis of V is changed to a new basis, and the new and the old bases are related by a matrix A, then the new coordinates of the same vector in V^* is related to the old coordinates with the same matrix A, while the new coordinates of the same vector in V is related to the old coordinates with the transpose of the inverse of A. The naming of the "contravariant" is due to the involvement of the inverse of matrix A.

The terms degree, rank and order are often used interchangeably in literature due to historical reasons. We adopt the term "degree", because the tensor space is a power of degree k (see remarks in Sec. 8 of Chap. 1).

In Chap. 5, we gave examples of tensor product spaces, with the space of matrices as $U \otimes V$, the space of bilinear forms as $U^* \otimes V^*$ and the space of linear mappings as $U \otimes V^*$. As special cases when $U = V$, they become examples of contravariant tensor space $V^{\otimes(2,0)}$, covariant tensor space $V^{\otimes(0,2)}$ and mixed tensor space $V^{\otimes(1,1)}$.

Definition 2. (Tensor spaces of higher degrees) The tensor product space

$$V^{\otimes(p,q)} \stackrel{\text{def}}{=} \underbrace{V \otimes \cdots \otimes V}_{p} \otimes \underbrace{V^* \otimes \cdots \otimes V^*}_{q}$$

is called the **tensor space** (or **tensor power space**) of type (p,q) over V, alternatively also denoted by $T^p_q(V)$. The vectors in $V^{\otimes(p,q)}$ are called **tensors** of type (p,q) over V. In particular, $V^{\otimes(p,0)}$ is called the **contravariant tensor space** of degree p. $V^{\otimes(0,q)}$ is called the **covariant tensor space** of degree q. Furthermore, $V^{\otimes(1,0)}$ is the same as V; $V^{\otimes(0,1)}$ is the same as V^*; $V^{\otimes(0,0)}$ is the same as F, which is the ground field of V.

§2. Change of Basis

We consider how the coordinates of tensors change if the basis of the underlying vector space changes. Starting now we will use a notational convention to distinguish the coordinates of contravariant tensors and covariant tensors. We use upper indices for contravariant coordinates and lower indices for covariant coordinates.

Let $\{\mathbf{e}_1, \ldots, \mathbf{e}_n\}$ be a basis of V. Suppose V undergoes a change of basis

$$\bar{\mathbf{e}}_i = \sum_{k=1}^{n} A^k{}_i \mathbf{e}_k, \tag{6.1}$$

where $A^k{}_i$ is the element at k^{th} row and i^{th} column of matrix A.

Theorem 1. (Change of coordinates for contravariant and covariant vectors) *Suppose the underlying vector space V undergoes a basis change and let A be the transition matrix.*

A contravariant vector $v \in V$ changes coordinates according to

$$\bar{v}^i = \sum_{k=1}^{n} [A^{-1}]^i{}_k v^k. \tag{6.2}$$

A covariant vector $u \in V^*$ changes coordinates according to

$$\bar{u}_i = \sum_{k=1}^{n} A^k{}_i u_k. \tag{6.3}$$

Remark. The naming of "contravariant" and "covariant" is with respect to the transition matrix A of basis transformation Eq. 6.1. The transformation of a covariant vector $\mathbf{u}$ in Eq. 6.3 involves the same matrix A, while the transformation of a contravariant vector $\mathbf{v}$ in Eq. 6.2 involves (the transpose of) the inverse of matrix A. If we call the transformation of basis with matrix A the "forward" transformation, then the transformation of the coordinates of vectors $\mathbf{v} \in V$ is the "backward" transformation, with an analogy: if one rides on the train and the train moves forward, the trees outside seem to move backward.

Theorem 2. (Change of coordinates for tensors) *Suppose the underlying vector space V undergoes a basis change and let A be the transition matrix.*

A contravariant tensor $\xi \in V^{\otimes(2,0)}$ changes coordinates according to

$$\bar{\xi}^{ij} = \sum_{k,l=1}^{n} [A^{-1}]^{i}{}_{k} [A^{-1}]^{j}{}_{l} \xi^{kl}. \tag{6.4}$$

A covariant tensor $\zeta \in V^{\otimes(0,2)}$ changes coordinates according to

$$\bar{\zeta}_{ij} = \sum_{k,l=1}^{n} A^{k}{}_{i} A^{l}{}_{j} \zeta_{kl}. \tag{6.5}$$

A mixed tensor $\eta \in V^{\otimes(1,1)}$ changes coordinates according to

$$\bar{\eta}^{i}{}_{j} = \sum_{k,l=1}^{n} [A^{-1}]^{i}{}_{k} A^{l}{}_{j} \eta^{k}{}_{l}. \tag{6.6}$$

§3. Induced Inner Product

Definition 3. (Induced inner product) Suppose the underlying space V is equipped with an inner product $\langle \cdot, \cdot \rangle$. There exists a unique inner product in $V^{\otimes(p,0)}$ satisfying

$$\langle \mathbf{u}_1 \otimes \cdots \otimes \mathbf{u}_p, \mathbf{v}_1 \otimes \cdots \otimes \mathbf{v}_p \rangle = \langle \mathbf{u}_1, \mathbf{v}_1 \rangle \cdots \langle \mathbf{u}_p, \mathbf{v}_p \rangle.$$

This inner product is called the **induced inner product** in $V^{\otimes(p,0)}$.

If $\{\mathbf{e}_1, \ldots, \mathbf{e}_n\}$ is an orthonormal basis for V with respect to $\langle \cdot, \cdot \rangle$, then $\{\mathbf{e}_{i_1} \otimes \cdots \otimes \mathbf{e}_{i_p} \mid i_1, \ldots, i_p = 1, \ldots, n\}$ forms an orthonormal basis for $V^{\otimes(p,0)}$.

§4. Lowering and Raising Indices—Isomorphisms

Let $(V, \langle \cdot, \cdot \rangle)$ be an inner product space. Let $\{\mathbf{e}_1, \ldots, \mathbf{e}_n\}$ be a basis for V and $g_{ij} = \langle \mathbf{e}_i, \mathbf{e}_j \rangle$. There exists an isomorphism $\Phi : V \to V^*$ such that a vector $\mathbf{v} = x^1 \mathbf{e}_1 + \cdots + x^n \mathbf{e}_n \in V$ is mapped to its metric dual $\mathbf{v}^* = x_1 f_1 + \cdots + x_n f_n$, and vice versa, where $\{f_1, \ldots, f_n\}$ is the affine dual basis. The coordinates of the metric dual $(x_1, \ldots, x_n)$ are related to $(x^1, \ldots, x^n)$ as follows,

$$x_i = \sum_{k=1}^{n} g_{ik} x^k, \tag{6.7}$$

$$x^i = \sum_{k=1}^{n} g^{ik} x_k, \tag{6.8}$$

where $[g^{ij}]$ is the inverse matrix of $[g_{ij}]$.

Now using the isomorphism $\Phi : V \to V^*$, we can define a linear mapping $\Pi^{\circ\circ}_{\downarrow} : T^2_0(V) \to T^1_1(V)$ such that

$$\mathbf{u} \otimes \mathbf{v} \mapsto \mathbf{u} \otimes \mathbf{v}^*.$$

The mapping $\Pi^{\circ\circ}_{\downarrow}$ is a linear isomorphism from $T^2_0(V)$ to $T^1_1(V)$, induced by the identity mapping $\mathrm{id} : V \to V; \mathbf{u} \mapsto \mathbf{u}$ and $\Phi : V \to V^*; \mathbf{v} \mapsto \mathbf{v}^*$. $\Pi^{\circ\circ}_{\downarrow}$ maps a contravariant tensor to a mixed tensor.

In coordinate notation, let w have components w^{ij} and $\Pi^{\circ\circ}_{\downarrow}(w)$ have components $w^i{}_j$. They are related by

$$w^i{}_j = \sum_{k=1}^{n} g_{kj} w^{ik}. \tag{6.9}$$

The inverse of $\Pi^{\circ\circ}_{\downarrow}$, denoted by $\Pi^{\circ\uparrow}_{\circ} : T^1_1(V) \to T^2_0(V)$ is an isomorphism from $T^1_1(V)$ to $T^2_0(V)$. In coordinate notation, we have

$$w^{ij} = \sum_{k=1}^{n} g^{ik} w^j{}_k. \tag{6.10}$$

The isomorphism $\Pi^{\circ\circ}_{\downarrow} : T^2_0(V) \to T^1_1(V)$ is called **lowering the index**, because in the coordinate expression Eq. 6.9 one index is lowered. The isomorphism $\Pi^{\circ\uparrow}_{\circ} : T^1_1(V) \to T^2_0(V)$ is called **raising the index** because one index is raised in the coordinate expression Eq. 6.10.

Since all the tensor spaces $T^2_0(V)$, $T^1_1(V)$ and $T^0_2(V)$ are isomorphic to each other, we can raise or lower more than one index. For example, lowering indices twice

$$w_{ij} = \sum_{k,l=1}^{n} g_{ik} g_{jl} w^{kl} \tag{6.11}$$

defines an isomorphism $\Pi^{\circ\circ}_{\downarrow\downarrow} : T^2_0(V) \to T^0_2(V)$. Raising indices twice

$$w^{ij} = \sum_{k,l=1}^{n} g^{ik} g^{jl} w_{kl} \tag{6.12}$$

establishes an isomorphism $\Pi^{\uparrow\uparrow}_{\circ\circ} : T^0_2(V) \to T^2_0(V)$.

In the case of orthonormal basis, $g_{ij} = \delta_{ij}$. The components of all different (p, q) types of tensor with equal $p + q$ are the same. For example, $w^{ij} = w^i_j = w_{ij}$ and we do not even need to distinguish the upper indices from the lower indices.

When we deal with higher degree tensor powers, the order of indices is important and we need to be specific about which index is raised or lowered to avoid ambiguity. In general we can have a tensor power of V and V^* in any order. For example,

$$w^i{}_j{}^{kl} = \sum_{k,l=1}^{n} g_{jp} g^{kq} w^{ip}{}_q{}^l \tag{6.13}$$

defines an isomorphism $\Pi^{\circ\circ\uparrow\circ}_{\downarrow\circ} : V \otimes V \otimes V^* \otimes V \to V \otimes V^* \otimes V \otimes V$.

Remark. Active View vs. Passive View:

The above discussion is the active view. That is to view the raising and lowering of indices as linear transformations (isomorphisms). In the passive view, the coordinates with raised or lowered indices are considered different coordinates of the same tensor under different bases. The same tensor w has coordinates in the form of w^{ijkl} under basis $\{\mathbf{e}_i \otimes \mathbf{e}_j \otimes \mathbf{e}_k \otimes \mathbf{e}_l\}^n_{i,j,k,l=1}$, but has coordinates $w^{ip}{}_q{}^l$ under basis $\{\mathbf{e}_i \otimes \mathbf{e}_p \otimes \hat{\mathbf{e}}_q \otimes \mathbf{e}_l\}^n_{i,p,q,l=1}$, and coordinates $w^i{}_j{}^{kl}$ under basis $\{\mathbf{e}_i \otimes \hat{\mathbf{e}}_j \otimes \mathbf{e}_k \otimes \mathbf{e}_l\}^n_{i,j,k,l=1}$, where $\{\hat{\mathbf{e}}_j\}^n_{j=1}$ is the reciprocal basis of $\{\mathbf{e}_i\}^n_{i=1}$.

Chapter 7

Tensor Algebra

Algebra is the name for a branch of mathematics. It is also overloaded to refer to some specific mathematical entities, such as tensor algebra, exterior algebra or Grassmann algebra, geometric algebra or Clifford algebra, Lie algebra, etc. The vectors in a vector space have two operations: the vector addition and scalar-vector multiplication. If the vectors are endowed with an additional structure—the multiplication of two vectors yielding another vector, then the system is called an algebra. Tensor algebra can have both meanings, but in this chapter, we discuss tensor algebra in the second sense.

§1. Tensor Product of Tensors

Let $x \in T_0^p(V)$ and $y \in T_0^r(V)$. Since $T_0^p(V)$ and $T_0^r(V)$ are vector spaces, we can form the tensor product space of them. $T_0^p(V) \otimes T_0^r(V)$ is isomorphic to $T_0^{p+r}(V)$. The mapping $\otimes : T_0^p(V) \times T_0^r(V) \to T_0^{p+r}(V)$ assigns $x = u_1 \otimes \cdots \otimes u_p$ and $y = v_1 \otimes \cdots \otimes v_r$ to

$$x \otimes y = u_1 \otimes \cdots \otimes u_p \otimes v_1 \otimes \cdots \otimes v_r.$$

Similarly we can define the tensor product of two covariant tensors. The tensor product $T_q^0(V) \otimes T_s^0(V)$ is isomorphic to $T_{q+s}^0(V)$.

In the most general case, the tensor product $T_q^p(V) \otimes T_s^r(V)$ is isomorphic to $T_{q+s}^{p+r}(V)$.

The tensor product of two vector spaces $U \otimes V$ is also called the "outer product" because $U \otimes V$ is a different vector space. It is "out of" U and it is "out of" V. Tensor spaces $T_0^p(V)$ of different degrees are different vector spaces and their vectors cannot be added together. The tensor product mapping $\otimes : T_0^p(V) \times T_0^r(V) \to T_0^{p+r}(V)$ is not a binary operation within one space. We really wish to make the mapping $\otimes$ a binary operation within one space though. To do so, we resort to the construction of direct sum of all these tensor power spaces $T_0^p(V)$, for $p = 0, 1, 2, \ldots$.

§2. Tensor Algebra

* **Review**: Linear Algebra—Definition of algebra

A vector space V over field F is called an **algebra** (or **linear algebra**) **over field F**, if V has another operation, called the vector multiplication, $(\times) : V \times V \to V$; $(\mathbf{u}, \mathbf{v}) \mapsto \mathbf{u} \times \mathbf{v}$, such that the following conditions are satisfied, for all $\mathbf{u}, \mathbf{v}, \mathbf{w} \in V$ and $a, b \in F$.

(1) Left distributive law: $(\mathbf{u} + \mathbf{v}) \times \mathbf{w} = \mathbf{u} \times \mathbf{w} + \mathbf{v} \times \mathbf{w}$.
(2) Right distributive law: $\mathbf{u} \times (\mathbf{v} + \mathbf{w}) = \mathbf{u} \times \mathbf{v} + \mathbf{u} \times \mathbf{w}$.
(3) $(a\mathbf{u}) \times (b\mathbf{v}) = (ab)(\mathbf{u} \times \mathbf{v})$.

Example 1. (Polynomials over field F as commutative algebra) All the polynomials $F[x]$ of a single variable x over a field F is an example of a commutative algebra over field F. The vector multiplication is the usual multiplication of polynomials.

Example 2. (Square matrices $M_{n,n}$ as associative algebra) All of the $n \times n$ matrices $M_{n,n}$ form a vector space with respect to matrix addition and scalar multiplication. Matrices also have multiplication defined. $M_{n,n}$ forms an algebra with respect to addition, scalar multiplication and matrix multiplication. This algebra is associative but not commutative.

Example 3. (Cross product algebra) Let $\mathbf{u} = (x_1, x_2, x_3), \mathbf{v} = (y_1, y_2, y_3) \in \mathbb{R}^3$. The cross product of $\mathbf{u} \times \mathbf{v}$ is defined to be a vector in $\mathbb{R}^3$ as

$$\mathbf{u} \times \mathbf{v} \overset{\text{def}}{=} (x_2 y_3 - x_3 y_2, x_3 y_1 - x_1 y_3, x_1 y_2 - x_2 y_1).$$

The vector space $\mathbb{R}^3$ with vector cross product as the vector multiplication is a nonassociative algebra over $\mathbb{R}$.

Example 4. (Lie algebra) Let V be an algebra over a field F and $(\times) : V \times V \to V$ is the vector multiplication. V is called a Lie algebra if the vector multiplication also satisfies the following conditions, for all $\mathbf{u}, \mathbf{v}, \mathbf{w} \in V$ and $a, b \in F$.

(1) Antisymmetry:
$$\mathbf{u} \times \mathbf{v} = -\mathbf{v} \times \mathbf{u}.$$

(2) Jacobi identity:
$$\mathbf{u} \times (\mathbf{v} \times \mathbf{w}) + \mathbf{v} \times (\mathbf{w} \times \mathbf{u}) + \mathbf{w} \times (\mathbf{u} \times \mathbf{v}) = 0.$$

The algebra $\mathbb{R}^3$ with vector cross product is a Lie algebra.

In Chap. 3, we defined the direct sum $U \oplus V$ of two vector spaces. We now define the direct sum of tensor spaces $T_q^p(V)$, $p, q = 0, 1, 2, \ldots$ to be

$$T(V) \overset{\text{def}}{=} \bigoplus_{p,q=0}^{\infty} T_q^p(V) = T_0^0(V) \oplus T_0^1(V) \oplus T_1^0(V) \oplus T_0^2(V) \oplus T_1^1(V) \oplus T_2^0(V) \oplus \cdots.$$

Then $\otimes$ becomes a binary operation $\otimes : T(V) \times T(V) \to T(V)$. $T(V)$ is an algebra over field F regarding the tensor multiplication $\otimes$. It is called the **tensor algebra** over V. Now tensors of different types can be added and the sum is understood as the direct sum. Each tensor space $T_q^p(V)$ is a linear subspace of $T(V)$. The product of a tensor in subspace $T_0^p(V)$ with a tensor in subspace $T_0^r(V)$ is a tensor in subspace $T_0^{p+r}(V)$. An algebra with properties like this is called a **graded algebra**.

§3. Contraction of Tensors

We start with an example. Let $w \in T_1^1(V)$ be a type $(1,1)$ tensor. Let $\Psi : V \times V^* \to \mathbb{R}$ such that for any $\mathbf{u} \in V$ and $\mathbf{v} \in V^*$, $\Psi(\mathbf{u}, \mathbf{v}) \overset{\text{def}}{=} \mathbf{v}(\mathbf{u})$, where $\mathbf{v}(\mathbf{u})$ is the action of linear function $\mathbf{v}$ on $\mathbf{u}$.

From the unique factorization property of tensor product, we know that there exists a unique linear mapping $C_1^1 : T_1^1(V) \to \mathbb{R}$ such that

$$C_1^1(\mathbf{u} \otimes \mathbf{v}) = \Psi(\mathbf{u}, \mathbf{v}) = \mathbf{v}(\mathbf{u}).$$

In the coordinate form, if w has coordinates w_j^i under a certain basis, then

$$C_1^1(w) = \sum_{k=1}^{n} w_k^k,$$

which is equal to the trace of the matrix $[w_j^i]$.

$C_1^1(w)$ so defined is called the **contraction** of tensor w with respect to (the 1st) upper index and (the 1st) lower index. This leads to the general definition of contraction of a tensor.

Definition. (Contraction of a tensor) Let $T_q^p(V)$ be a tensor space of type (p, q). For any $1 \le s \le p$ and $1 \le t \le q$, from the unique factorization property of the tensor product, there exists a unique linear mapping $C_t^s : T_q^p(V) \to T_{q-1}^{p-1}(V)$ such that for any $\mathbf{u}_1, ..., \mathbf{u}_p \in V$ and $\mathbf{v}_1, ..., \mathbf{v}_q \in V^*$,

$$C_t^s(\mathbf{u}_1 \otimes \cdots \otimes \mathbf{u}_s \otimes \cdots \otimes \mathbf{u}_p \otimes \mathbf{v}_1 \otimes \cdots \otimes \mathbf{v}_t \otimes \cdots \otimes \mathbf{v}_q)$$
$$= [\mathbf{v}_t(\mathbf{u}_s)]\, \mathbf{u}_1 \otimes \cdots \otimes \mathbf{u}_{s-1} \otimes \mathbf{u}_{s+1} \cdots \otimes \mathbf{u}_p \otimes \mathbf{v}_1 \otimes \cdots \otimes \mathbf{v}_{t-1} \otimes \mathbf{v}_{t+1} \cdots \otimes$$

$\mathbf{v}_q$, where $\mathbf{v}_t(\mathbf{u}_s)$ is the action of linear function $\mathbf{v}_t$ on vector $\mathbf{u}_s$. The mapping $C_t^s : T_q^p(V) \to T_{q-1}^{p-1}(V)$ is called the **contraction** of a tensor of type (p,q) with respect to the sth upper index and t^{th} lower index.

The contraction of a tensor of type (p, q) results in a tensor of type $(p-1, q-1)$. It can be viewed as a single argument operator in the tensor space $T(V)$, namely $C_t^s : T(V) \to T(V)$.

Theorem. *Let $w \in T_q^p(V)$. Then $C_t^s(w)$ can be obtained using coordinates by identifying the sth contravariant index with the tth covariant index and sum over them. Namely,*

$$C_t^s(w) = \sum_{k=1}^{n} w^{\cdots k \cdots}_{\cdots k \cdots}.$$

Chapter 8

Dynamics: The Inertia Tensor

Oftentimes tensors appear in some *moments*, like the moment of inertia tensor and the electric and magnetic multipole moment tensors. Traditional physics textbooks define the moment of inertia tensor as a matrix but give it a surname Tensor. Logically a term so defined should be called inertia matrix. Why is it called inertia tensor? This puzzles many students. We show in this chapter that this entity should be defined as a linear transformation, or a quadratic form. Linear transformations and quadratic forms are models of tensors. What is essential is that it is not just a matrix. A matrix is only the representation of a linear transformation or quadratic form, and the representation matrix should change when the basis of the space changes.

In the following discussions, we assume a Cartesian coordinate system with the chosen origin O. The position P of any mass particle in space is represented by the radius vector $\mathbf{r} = \overrightarrow{OP}$.

§1. Angular Momentum

We first consider a single mass particle with mass m and location $\mathbf{r}$. Its velocity is defined to be the time derivative of $\mathbf{r}$,

$$\mathbf{v} \overset{\text{def}}{=} \dot{\mathbf{r}} \overset{\text{def}}{=} \frac{d\mathbf{r}}{dt}.$$

Its momentum is defined to be

$$\mathbf{P} \overset{\text{def}}{=} m\mathbf{v}.$$

Suppose a force $\mathbf{F}$ is acting on the particle. By Newton's second law, we have

$$\mathbf{F} = \frac{d\mathbf{P}}{dt}.$$

Here $\mathbf{F}$ is understood as the sum of all forces acting on this particle. If the force $\mathbf{F} = 0$, the momentum $\mathbf{P}$ is conserved.

Definition 1. (Kinetic energy of a particle) The kinetic energy of the particle is defined to be $T \overset{\text{def}}{=} \frac{1}{2}m\mathbf{v}^2$.

Note, the kinetic energy is independent of the choice of origin O.
The motion can also be described from an angular perspective.

Definition 2. (Angular momentum and torque of a particle)
The **angular momentum** of the particle **with respect to point** O is

$$\mathbf{L} \overset{\text{def}}{=} \mathbf{r} \times \mathbf{P} \overset{\text{def}}{=} m\mathbf{r} \times \mathbf{v}. \tag{8.1}$$

The **torque of the force** $\mathbf{F}$ **with respect to point** O is

$$\boldsymbol{\tau} \overset{\text{def}}{=} \mathbf{r} \times \mathbf{F}. \tag{8.2}$$

It can be easily deduced that

$$\boldsymbol{\tau} = \frac{d\mathbf{L}}{dt}. \tag{8.3}$$

If $\boldsymbol{\tau} = 0$, then the angular momentum is conserved.

There are two special cases in which the angular momentum is conserved.

Case (1) zero force:

If $\mathbf{F} = 0$, then of course $\boldsymbol{\tau} = 0$, hence $\mathbf{L}$ is a constant. The motion is on a straight line.

Case (2) central force:

If the force goes through O, then $\boldsymbol{\tau} = \mathbf{r} \times \mathbf{F} = 0$. This is the case of central force, and the angular momentum $\mathbf{L}$ is conserved. In such a case, the motion of the particle is in a plane orthogonal to $\mathbf{L}$. Within the plane, we can use the polar coordinates (r, θ) to describe the motion. We can define the angular velocity

$$\omega \stackrel{\text{def}}{=} \dot{\theta} \stackrel{\text{def}}{=} \frac{d\theta}{dt}.$$

We can make angular velocity $\boldsymbol{\omega}$ a vector by giving it a direction, the direction orthogonal to the plane of motion, which is the same direction of $\mathbf{L}$. This way we have

$$\mathbf{v} = \boldsymbol{\omega} \times \mathbf{r}.$$

Note the angular description of the motion of the particle applies only to the case of planar motion.

Suppose we have a system of N particles, each having mass m_δ and position $\mathbf{r}_\delta$, $\delta = 1, \ldots, N$. The above definition of angular momentum, kinetic energy, and torque apply to each of them.

$$\mathbf{L}_\delta \stackrel{\text{def}}{=} m_\delta \mathbf{r}_\delta \times \mathbf{v}_\delta,$$

$$T_\delta \stackrel{\text{def}}{=} \frac{1}{2} m_\delta \mathbf{v}_\delta^2,$$

$$\boldsymbol{\tau}_\delta \stackrel{\text{def}}{=} \mathbf{r}_\delta \times \mathbf{F}_\delta, \quad \delta = 1, \ldots, N.$$

We define the total momentum, total angular momentum, total kinetic energy, total force and total torque of the system as the sum over all particles as follows.

Definition 3. (Total quantities of a system) The **total momentum** of the system is the sum

$$\mathbf{P} \stackrel{\text{def}}{=} \sum_{\delta=1}^{N} \mathbf{P}_\delta \stackrel{\text{def}}{=} \sum_{\delta=1}^{N} m_\delta \mathbf{v}_\delta. \tag{8.4}$$

The **total angular momentum** of the system with respect to point O is the sum

$$\mathbf{L} \overset{\text{def}}{=} \sum_{\delta=1}^{N} \mathbf{L}_\delta \overset{\text{def}}{=} \sum_{\delta=1}^{N} \left(m_\delta \mathbf{r}_\delta \times \mathbf{v}_\delta \right). \tag{8.5}$$

The **total kinetic energy** is the sum

$$T \overset{\text{def}}{=} \sum_{\delta=1}^{N} T_\delta \overset{\text{def}}{=} \sum_{\delta=1}^{N} \frac{1}{2} m_\delta \mathbf{v}_\delta^2. \tag{8.6}$$

The **total force** is the sum

$$\mathbf{F} \overset{\text{def}}{=} \sum_{\delta=1}^{N} \mathbf{F}_\delta. \tag{8.7}$$

The **total torque** with respect to point O is the sum

$$\boldsymbol{\tau} \overset{\text{def}}{=} \sum_{\delta=1}^{N} \boldsymbol{\tau}_\delta \overset{\text{def}}{=} \sum_{\delta=1}^{N} \left(\mathbf{r}_\delta \times \mathbf{F}_\delta \right). \tag{8.8}$$

Remark. Caution: it needs to be emphasized that the angular momentum $\mathbf{L}_\delta$ so defined is not an absolute quantity. It is relative to the chosen origin O. If a rigid body moves around a fixed axis, then it has more than one fixed point. Selecting different points as the origin, the angular momentum $\mathbf{L}_\delta$ would have different values. The total angular momentum $\mathbf{L}$ and the total torque $\boldsymbol{\tau}$ of the system are also relative to the point O. The kinetic energy T does not depend on the choice of origin though.

We can deduce the dynamical laws of motion for a system of particles.

$$\mathbf{F} = \frac{d\mathbf{P}}{dt}, \ \boldsymbol{\tau} = \frac{d\mathbf{L}}{dt}.$$

As corollaries, if $\mathbf{F} = 0$, $\mathbf{P}$ is conserved; if $\boldsymbol{\tau} = 0$, $\mathbf{L}$ is conserved.

§2. Rotation of Rigid Body around a Fixed Point

The above definitions of the total angular momentum and total kinetic energy apply to general systems. They apply to rigid bodies too, as a special case. A rigid body is a system in which any two mass elements keep a constant distance during motion. For simplicity, we first consider a rigid body with N discrete point particles with mass $m_\delta, \delta = 1, \ldots, N$. We use δ for the index of particles while we reserve $i, j = 1, 2, 3$ for indices of spatial directions x, y, z. We can think of that any two mass elements are connected by an extremely light (massless) thin rod to enforce the constant distance constraint. We call each particle a mass element. It will be an easy task to change from discrete mass distribution to continuous mass distribution where each mass element dm is infinitesimal. In such a case, the summation is simply replaced by integration (the Riemann sum).

In this section we discuss the motion of a rigid body around a fixed point O, which we choose to be the origin of the coordinate system. It does not have to be the center of mass of the body. All the definitions of the total momentum (Eq. 8.4), angular momentum (Eq. 8.5) and kinetic energy (Eq. 8.6) of a system with respect to point O apply to a rigid body. There is something special for a rigid body. That is, all the mass elements have a common angular velocity $\boldsymbol{\omega}$.

Theorem 1. *Suppose a rigid body undergoes motion around a fixed point O. At any time instance t, there exists a line l through O on which the material points of the body have zero instantaneous velocity. This line is called the instantaneous axis of rotation at time t. All other material points of the body rotate around l with angular velocity $\boldsymbol{\omega}$ at this time instance. The velocity $\mathbf{v}_\delta$ of mass element m_δ is related to $\boldsymbol{\omega}$ by*

$$\mathbf{v}_\delta = \boldsymbol{\omega} \times \mathbf{r}_\delta, \ \delta = 1, \ldots, N.$$

First, we look at the angular momentum of mass element m_δ with respect to the point O.

$$\mathbf{L}_\delta \overset{\text{def}}{=} m_\delta \mathbf{r}_\delta \times \mathbf{v}_\delta$$
$$= m_\delta \mathbf{r}_\delta \times (\boldsymbol{\omega} \times \mathbf{r}_\delta)$$
$$= m_\delta \left\{ \left(\mathbf{r}_\delta^2 \right) \boldsymbol{\omega} - \mathbf{r}_\delta \left(\mathbf{r}_\delta \cdot \boldsymbol{\omega} \right) \right\}.$$

It is easy to see that the angular momentum $\mathbf{L}_\delta$ depends on $\boldsymbol{\omega}$ linearly, but in general $\mathbf{L}_\delta$ is not in the same direction as $\boldsymbol{\omega}$, because $\mathbf{L}_\delta$ is a linear combination of $\boldsymbol{\omega}$ and $\mathbf{r}_\delta$.

A student may have learned in classroom that the angular momentum $\mathbf{L}$ is not in the same direction of the angular velocity $\boldsymbol{\omega}$ for a rigid body rotating around a fixed point. Note this is not peculiar to rigid bodies, nor is it peculiar to rotation around a fixed point. It is true for a single mass particle. This is just due to the definition of angular momentum (with respect to a point). In the situation of a rigid body, all the mass elements have a common angular velocity $\boldsymbol{\omega}$ and the total angular momentum $\mathbf{L}$ is the sum of $\mathbf{L}_\delta$ over all the mass elements. It is a misconception that the angular momentum $\mathbf{L}$ is in the same direction as angular velocity $\boldsymbol{\omega}$ for a rigid body rotating around a fixed axis. This misconception will be further analyzed in Sec. 3.

We define a linear mapping $\mathscr{I}_\delta : \mathbb{R}^3 \to \mathbb{R}^3$, such that for any $\boldsymbol{\omega} \in \mathbb{R}^3$,

$$\boldsymbol{\omega} \mapsto \mathscr{I}_\delta \boldsymbol{\omega} \overset{\text{def}}{=} \mathbf{L}_\delta.$$

$\mathscr{I}_\delta$ is called the **moment of inertia operator** (or just the **inertia operator**) defined for particle δ **with respect to point** O.

Note $\mathscr{I}_\delta$ is a linear mapping defined for the single particle δ. $\boldsymbol{\omega}$ is the input while $\mathbf{L}_\delta$ is the output. The particle information—the mass m_δ and position $\mathbf{r}_\delta$ are parameters to define this linear mapping $\mathscr{I}_\delta$. Since a linear mapping is a tensor, we can call it the **inertia tensor** for particle δ with respect to O. We can further rewrite this inertia operator as

$$\mathscr{I}_\delta = m_\delta \left[(\mathbf{r}_\delta \cdot \mathbf{r}_\delta)\, \mathscr{E} - \mathbf{r}_\delta \otimes \mathbf{r}_\delta \right],$$

where $\mathscr{E}$ is the identity operator with the property $\mathscr{E}\boldsymbol{\omega} = \boldsymbol{\omega}$. The first $\mathbf{r}_\delta$ in $\mathbf{r}_\delta \otimes \mathbf{r}_\delta$ is a vector in space $V = \mathbb{R}^3$ while the second $\mathbf{r}_\delta$ is viewed as a linear function, which is in V^*.

Any linear mapping is a tensor, no matter it is the tensor product of two vectors or not. If it is, then it is a decomposable tensor (see Sec. 5 of Chap. 5). Now $\mathscr{I}_\delta$ is the sum of two parts. The second part $\mathbf{r}_\delta \otimes \mathbf{r}_\delta$ is a decomposable tensor, but the first part is a multiple of $\mathscr{E}$, which is not decomposable. In fact $\mathscr{I}_\delta$ is not decomposable (the proof is left to the reader as exercise).

Because all mass elements have a common angular velocity $\boldsymbol{\omega}$, the total angular momentum also depends on $\boldsymbol{\omega}$ linearly:

$$\mathbf{L} = \sum_{\delta=1}^{N} \mathbf{L}_\delta = \left(\sum_{\delta=1}^{N} \mathscr{I}_\delta \right) \boldsymbol{\omega}.$$

Definition 4. (Inertia operator) The **inertia operator of the rigid body with respect to point** O is defined to be $\mathscr{I} : \mathbb{R}^3 \to \mathbb{R}^3$,

$$\mathscr{I} \overset{\text{def}}{=} \sum_{\delta=1}^{N} \mathscr{I}_\delta \overset{\text{def}}{=} \sum_{\delta=1}^{N} m_\delta \left[(\mathbf{r}_\delta \cdot \mathbf{r}_\delta) \, \mathscr{E} - \mathbf{r}_\delta \otimes \mathbf{r}_\delta \right].$$

Then the total angular momentum of the rigid body is

$$\mathbf{L} = \mathscr{I}\boldsymbol{\omega}.$$

$\mathscr{I}$ is the sum of N linear mappings. Therefore $\mathscr{I}$ itself is a linear mapping that maps the angular velocity to the total angular momentum of the rigid body. In tensor language, it is the sum of N tensors. Therefore itself is a tensor. Is it a decomposable tensor? This depends on the mass configuration. In general it is not decomposable, but for special cases, it may be.

The inertia tensor $\mathscr{I}$ has a matrix representation,

$$[\mathscr{I}] = \begin{bmatrix} I_{11} & I_{12} & I_{13} \\ I_{21} & I_{22} & I_{23} \\ I_{31} & I_{32} & I_{33} \end{bmatrix},$$

with

$$I_{11} = \sum_{\delta=1}^{N} m_\delta \left(y_\delta^2 + z_\delta^2 \right), \quad I_{12} = I_{21} = -\sum_{\delta=1}^{N} m_\delta x_\delta y_\delta,$$

$$I_{22} = \sum_{\delta=1}^{N} m_\delta \left(z_\delta^2 + x_\delta^2 \right), \quad I_{13} = I_{31} = -\sum_{\delta=1}^{N} m_\delta z_\delta x_\delta,$$

$$I_{33} = \sum_{\delta=1}^{N} m_\delta \left(x_\delta^2 + y_\delta^2 \right), \quad I_{23} = I_{32} = -\sum_{\delta=1}^{N} m_\delta y_\delta z_\delta.$$

In the case of continuous mass distribution,

$$\mathscr{I} = \int \left(\mathbf{r}^2 \mathscr{E} - \mathbf{r} \otimes \mathbf{r} \right) dm.$$

$\mathscr{I}$ is the sum (Riemann sum) of infinitely many infinitesimal linear transformations.

$$I_{11} = \int \left(y^2 + z^2\right) dm, \quad I_{12} = I_{21} = -\int xy\, dm,$$

$$I_{22} = \int \left(z^2 + x^2\right) dm, \quad I_{13} = I_{31} = -\int zx\, dm,$$

$$I_{33} = \int \left(x^2 + y^2\right) dm, \quad I_{23} = I_{32} = -\int yz\, dm.$$

Remark. Caution: do not forget the phrase "with respect to point O" in the definition, because it can be different with respect to different points. Oftentimes we omit this phrase with an implicit understanding that it is with respect to the fixed point in the case of rotation around a fixed point.

The kinetic energy of the mass element m_δ is

$$T_\delta = \frac{1}{2} m_\delta \mathbf{v}_\delta^2 = \frac{1}{2} m_\delta \mathbf{v}_\delta \cdot (\boldsymbol{\omega} \times \mathbf{r}_\delta) = \frac{1}{2} m_\delta \boldsymbol{\omega} \cdot (\mathbf{r}_\delta \times \mathbf{v}_\delta)$$
$$= \frac{1}{2} \boldsymbol{\omega} \cdot \mathbf{L}_\delta = \frac{1}{2} \boldsymbol{\omega} \cdot (\mathscr{I}_\delta \boldsymbol{\omega}).$$

This is a quadratic form acting on $\boldsymbol{\omega}$. Therefore the same inertia operator can be identified as a quadratic form $\mathscr{I}_\delta : \mathbb{R}^3 \to \mathbb{R}$. The result of the quadratic form $\mathscr{I}_\delta$ operating on $\boldsymbol{\omega}$ can be written as $\boldsymbol{\omega} \cdot \mathscr{I}_\delta \boldsymbol{\omega}$ or $\boldsymbol{\omega} \cdot \mathscr{I}_\delta \cdot \boldsymbol{\omega}$.
The sum

$$\mathscr{I} \stackrel{\text{def}}{=} \sum_{\delta=1}^{N} \mathscr{I}_\delta \stackrel{\text{def}}{=} \sum_{\delta=1}^{N} m_\delta \left[\left(\mathbf{r}_\delta^2\right) \langle \cdot, \cdot \rangle - \mathbf{r}_\delta \otimes \mathbf{r}_\delta \right]$$
$$= \left(\sum_{\delta=1}^{N} m_\delta \mathbf{r}_\delta^2 \right) \langle \cdot, \cdot \rangle - \sum_{\delta=1}^{N} m_\delta \mathbf{r}_\delta \otimes \mathbf{r}_\delta$$

is also a quadratic form, where $\langle \cdot, \cdot \rangle$ is the inner product and the dots are place holders for the inputs. When $\mathscr{I}$ is applied on a vector $\boldsymbol{\omega}$, both dots in $\langle \cdot, \cdot \rangle$ should be replaced by $\boldsymbol{\omega}$. Both $\mathbf{r}_\delta$ in $\mathbf{r}_\delta \otimes \mathbf{r}_\delta$ are considered linear functions in V^*.

When $\mathscr{I}$ is understood as a quadratic form, the total kinetic energy of the rigid body is written as

$$T = \frac{1}{2}\mathscr{I}(\boldsymbol{\omega}).$$

Let $\mathbf{n}$ be the unit vector in the direction of $\boldsymbol{\omega}$ so that the vector $\boldsymbol{\omega}$ can be written as $\boldsymbol{\omega} = \omega\mathbf{n}$.

Definition 5. (Scalar moment of inertia with respective to an axis) We define

$$I(\mathbf{n}) \stackrel{\text{def}}{=} \mathscr{I}(\mathbf{n}) \stackrel{\text{def}}{=} \mathbf{n} \cdot \mathscr{I}\mathbf{n},$$

and call it the **scalar moment of inertia with respect to axis n.**

When the axis $\mathbf{n}$ is implicitly understood, we can just write I instead of $I(\mathbf{n})$. It is easy to show that

$$T = \frac{1}{2}I\omega^2. \tag{8.9}$$

Theorem 2. *Let $\mathbf{r}_\delta^\perp$ be the distance of the mass element m_δ to the axis* $\mathbf{n}$. *Then*

$$I(\mathbf{n}) = \sum_{\delta=1}^{N} m_\delta \left(\mathbf{r}_\delta^\perp\right)^2. \tag{8.10}$$

Eq. 8.10 is often treated as the definition of scalar moment of inertia with respect to an axis in general physics texts.

§3. Rotation of Rigid Body around a Fixed Axis

We consider a rigid body that undergoes rotation around a fixed axis, whose direction is represented by a unit vector $\mathbf{n}$. The angular velocity $\boldsymbol{\omega}$ is in the direction of the rotation axis. It is convenient to choose this rotation axis as our z-axis.

This is a special case of motion about a fixed point. All discussions in the previous section apply to the case of rotation about a fixed axis.

For rotation about a fixed point, we have seen that

$$\mathbf{L} = \mathscr{I}\boldsymbol{\omega},$$

where $\mathscr{I}$ is the inertia operator. This means that the angular momentum depends on the angular velocity linearly, but they are not in the same direction in general. In the case of rotation about a fixed axis, there is a common misconception, which is rather subtle.

> *** Misconception**[1]
>
> In the rotation of a rigid body about a fixed axis, the angular momentum $\mathbf{L}$ is in the same direction as angular velocity $\boldsymbol{\omega}$. That is, $\mathbf{L} = I\boldsymbol{\omega}$, where I is the scalar moment of inertia relative to axis $\mathbf{n}$ as in Eq. 8.10.

A wrong analogy with the stress tensor might contribute in part to this misconception. This misconception states: for the special case of rotation around a fixed axis, $\mathbf{L}$ is in the same direction of $\boldsymbol{\omega}$ and they are related by a scalar factor $\mathbf{L} = I\boldsymbol{\omega}$, while in the general case of rotation around a fixed point, $\mathbf{L}$ and $\boldsymbol{\omega}$ are not in the same direction and they are related by a tensor in the form $\mathbf{L} = \mathscr{I}\boldsymbol{\omega}$. Compare this with the stress in liquids and solids. In a liquid the force $\mathbf{F}$ is in the same direction of the surface normal $\mathbf{S}$ and they are related by a scalar factor $\mathbf{F} = \sigma\mathbf{S}$, while in the case of solids, they are not in the same direction and are related by a tensor $\mathbf{F} = \Sigma\mathbf{S}$ (see Sec. 4 in Chap. 1).

This analogy does not go through. The above statements about the stress forces are correct, while the statements about the angular momentum are wrong. In either case, rotation around a fixed point or around a fixed axis, $\mathbf{L}$ and $\boldsymbol{\omega}$ are not in the same direction. The claim $\mathbf{L} = I\boldsymbol{\omega}$ is incorrect for rotation around a fixed axis, where I is a scalar and $\mathbf{L}$ is the angular momentum of the body relative to point O as defined in Definition 3. This error is easy to see. In the case of rotation around a fix axis, even if it is true that all the points on the rotation axis are fixed points during rotation, the angular momentum $\mathbf{L}$ is defined with respect to a specific point O on the axis $\mathbf{n}$. $\mathbf{L}$ is different with respect to a different point O' on the same axis $\mathbf{n}$. In general $\mathbf{L}$ with respect to point O so defined is not in the direction of the rotation axis $\mathbf{n}$, or angular velocity $\boldsymbol{\omega}$.

However, if we define a different *angular momentum* $\overline{\mathbf{L}}$ *relative to the axis* $\mathbf{n}$ as follows, we can have a simple relationship

$$\overline{\mathbf{L}} = I\boldsymbol{\omega},$$

where I is the scalar moment of inertia relative to axis $\mathbf{n}$.

---

[1]Neuenschwander, D. E. (2015). *Tensor Calculus for Physics*, p. 35, (Johns Hopkins University Press).

Definition 6. (Angular momentum and torque relative to an axis)
The **angular momentum** of mass element m_δ **with respect to axis n** is

$$\overline{\mathbf{L}}_\delta \overset{\text{def}}{=} m_\delta \mathbf{r}_\delta^\perp \times \mathbf{v}_\delta.$$

The **total angular momentum** of the rigid body **with respect to axis n** is
the sum

$$\overline{\mathbf{L}} \overset{\text{def}}{=} \sum_{\delta=1}^{N} \overline{\mathbf{L}}_\delta \overset{\text{def}}{=} \sum_{\delta=1}^{N} \left(m_\delta \mathbf{r}_\delta^\perp \times \mathbf{v}_\delta \right).$$

Let $\mathbf{F}_\delta$ be the force applied on mass element m_δ. We can decompose the
force : $\mathbf{F}_\delta = \mathbf{F}_\delta^\perp + \mathbf{F}_\delta^\parallel$, where $\mathbf{F}_\delta^\perp$ is perpendicular to $\mathbf{n}$ and $\mathbf{F}_\delta^\parallel$ is parallel
to $\mathbf{n}$. Define the **torque of $\mathbf{F}_\delta$ with respect to axis n** as

$$\overline{\boldsymbol{\tau}}_\delta \overset{\text{def}}{=} \mathbf{r}_\delta^\perp \times \mathbf{F}_\delta^\perp,$$

and the total torque relative to axis $\mathbf{n}$ to be

$$\overline{\boldsymbol{\tau}} \overset{\text{def}}{=} \sum_{\delta=1}^{N} \overline{\boldsymbol{\tau}}_\delta.$$

We will state without further explanation, the results of rotation about
a fixed axis. It can be proved that

$$\overline{\mathbf{L}} = I\boldsymbol{\omega}, \text{ and } T = \frac{1}{2} I \boldsymbol{\omega}^2.$$

We have the dynamic laws in the form

$$\overline{\boldsymbol{\tau}} = I \frac{d\boldsymbol{\omega}}{dt}, \text{ or } \overline{\boldsymbol{\tau}} = \frac{d\overline{\mathbf{L}}}{dt}.$$

If $\overline{\boldsymbol{\tau}} = 0$, both $\boldsymbol{\omega}$ and $\overline{\mathbf{L}}$ are conserved.

What are the relationships between $\overline{\mathbf{L}}$ and $\mathbf{L}$, and between $\overline{\boldsymbol{\tau}}$ and $\boldsymbol{\tau}$,
namely the quantities with respect to a point and those with respect to an
axis?

It is easy to find that $\overline{\mathbf{L}}$ and $\overline{\boldsymbol{\tau}}$ are exactly the z components of $\mathbf{L}$ and
$\boldsymbol{\tau}$. Namely,

$$\overline{\mathbf{L}} = \mathbf{L}_z, \ \overline{\boldsymbol{\tau}} = \boldsymbol{\tau}_z.$$

Therefore there is a difference in the case for $\mathbf{L}$ and $\boldsymbol{\tau}$, but no difference for
T, because the kinetic energy is not defined with respect to a point, or to
an axis.

It is often the case that graduate texts discuss rotation around a fixed point, while undergraduate texts discuss rotation around a fixed axis, which is simpler. However, some undergraduate texts fall into pitfalls regarding the angular momentum of a rigid body rotating around a fixed axis.

(1) Vague definitions of torque and angular momentum and implicit assumptions

In these texts, the torque and the angular momentum are not clearly defined, whether relative to an axis, or relative to a point. They may have an implicit assumption that the force $\mathbf{F}_\delta$ is perpendicular to the axis without explicitly stating it.

Schaum's outline book[2] just writes $\mathbf{L} = I\boldsymbol{\omega}$, without clearly defining angular momentum $\mathbf{L}$. Is this equation a definition, or a theorem? This is the first appearance of angular momentum $\mathbf{L}$ in the book.

(2) Some books define angular momentum $\mathbf{L}$ relative to a point clearly (as in Definition 3), but still draw a conclusion $\mathbf{L} = I\boldsymbol{\omega}$. This is a mistake then.[3]

We have seen that even if for the case of rotation around a fixed axis, we can only conclude

$$\overline{\mathbf{L}} = \mathbf{L}_z = I\boldsymbol{\omega},$$

but in general,

$$\mathbf{L} \neq I\boldsymbol{\omega}.$$

Question. For rotation around a fixed axis z, what are some sufficient conditions for $\mathbf{L} = I\boldsymbol{\omega}$? Namely, for $\mathbf{L}_x = \mathbf{L}_y = 0$.

Sufficient condition (1): if all the material points lie in one plane perpendicular to $\boldsymbol{\omega}$, then $\mathbf{L}_x = \mathbf{L}_y = 0$ and $\mathbf{L} = I\boldsymbol{\omega}$.

Sufficient condition (2): if the rigid body is symmetric about z axis, then $\mathbf{L}_x = \mathbf{L}_y = 0$ and $\mathbf{L} = I\boldsymbol{\omega}$.

Note, when we say the rigid body is symmetric about z axis, we mean if there is a mass element m at position (x, y, z), then there is also an equal mass m at position $(-x, -y, z)$. In such a case, the $\mathbf{L}_x$ and $\mathbf{L}_y$ components of these two symmetric mass elements cancel each other.

Cylindrical symmetry is a special case of such axial symmetry, and hence is a stronger sufficient condition. In general, an axial symmetry does not have to be cylindrical symmetry.

---

[2] Hecht, E. (2012). *Schaum's Outline of College Physics*, 11th ed., p. 129, (McGraw-Hill).

[3] Bauer, W. and Westfall, G. (2011). *University Physics with Modern Physics*, p. 337, (McGraw-Hill).

Question. For rotation around a fixed axis, what is a sufficient and necessary condition for $\mathbf{L}_x = \mathbf{L}_y = 0$ and $\mathbf{L} = I\boldsymbol{\omega}$?

Let us write the equation in coordinate form.

$$L_i = \sum_{j=1}^{3} I_{ij}\omega_j,$$

with $i = 1, 2, 3$ representing x, y, z directions. We have $\omega_x = \omega_y = 0$ and $\omega_z = \omega$. A sufficient and necessary condition for $\mathbf{L}_x = \mathbf{L}_y = 0$ and $\mathbf{L} = I\omega$ is $I_{13} = 0$ and $I_{23} = 0$. Namely,

$$\sum_{\delta=1}^{N} m_\delta x_\delta z_\delta = 0,$$

$$\sum_{\delta=1}^{N} m_\delta y_\delta z_\delta = 0.$$

Next let us analyze the total angular momentum $\mathbf{L}$ in more detail in the rotation around a fixed axis.

For a rigid body moving around a fixed point, we have

$$\boldsymbol{\tau} = \frac{d\mathbf{L}}{dt},$$

where both $\boldsymbol{\tau}$ and $\mathbf{L}$ are relative to point O. This is also true for rotation around a fixed axis.

If the force $\mathbf{F}_\delta$ is zero on all mass points except at point O, the rigid body is said to undergo free rotation around a fixed point. In this case, $\boldsymbol{\tau} \equiv 0$ and the angular momentum $\mathbf{L}$ is conserved. This does not imply that the angular velocity is a constant, because $\mathbf{L} \neq I\boldsymbol{\omega}$ in general. The angular velocity may change magnitude and direction. The rigid body may undergo precession and nutation.

For motion around a fixed axis, forces (applied on at least two points) are necessary to make the axis fixed. The simplest case is that there are two bearings on the axis, at point O and another point A. We suppose that all forces on the rigid body are applied on either point O or A and forces on other points are zero. We may call this free rotation about a fixed axis. In such a case, the torque from the force on A relative to point O is not zero. The z component of angular momentum L_z is conserved. The angular velocity is a constant. But the total angular momentum $\mathbf{L}$ is not a constant. As a result, L_x and L_y may change with time in general, except for some special cases that we have analyzed. For example, if the rigid body has an axial symmetry relative to the rotation axis, then $L_x = L_y = 0$. The total angular momentum $\mathbf{L} = \mathbf{L}_z$ is also conserved. Using the dynamical laws, we can solve for the constraining force $\mathbf{F}_A$ through bearing A. In the case when the rigid body does not have an axial symmetry, there is non-zero force $\mathbf{F}_A$ which may incur reaction force on the bearing, which in most cases are undesirable in mechanical engineering designs.

§4. Parallel Axis Theorem and Perpendicular Axis Theorem

When the rigid body has continuous mass distribution, all the summation over δ should be changed to integration of mass element $dm = \rho\, dx dy dz$, where ρ is the mass density. By Theorem 2, the scalar moment of inertia of a rigid body with respect to an axis is

$$I = \int \left(\mathbf{r}^{\perp}\right)^2 dm = \iiint \rho \left(\mathbf{r}^{\perp}\right)^2 dx dy dz. \qquad (8.11)$$

Take a solid cube for example. Suppose the cube has side length a and mass m. We choose the origin O to be the center of the cube, and axes x, y, z parallel to its three families of edges. Its scalar moment of inertia with respect to the z axis can be calculated straightforwardly to be

$$I_z = \frac{1}{6}ma^2.$$

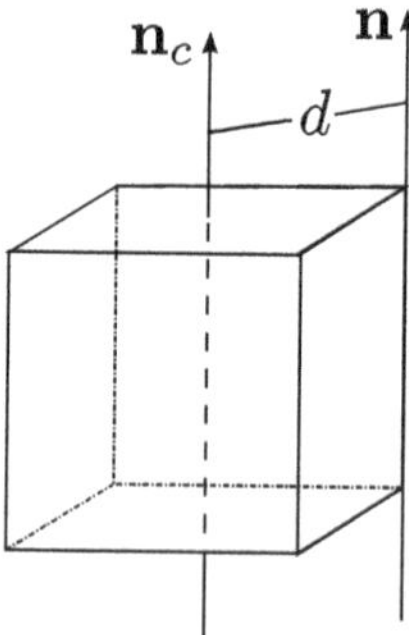

Figure 8.1   Parallel axis theorem

Theorem 3. (Parallel Axis Theorem of Steiner) *Let $I(\mathbf{n})$ and $I(\mathbf{n}_c)$ be the momenta of inertia of the same rigid body with respect to axes $\mathbf{n}$ and $\mathbf{n}_c$, where $\mathbf{n}$ and $\mathbf{n}_c$ are parallel and $\mathbf{n}_c$ goes through the center of mass of the body. Then*

$$I(\mathbf{n}) = I(\mathbf{n}_c) + md^2,$$

where m is the mass of the body and d is the distance between the two axes (Figure 8.1).

This theorem can help us calculate some moment of inertia faster. For example, using Steiner's theorem, we can find easily that the moment of inertia of a solid cube with respect to one of its edges to be (Figure 8.1):

$$I(\mathbf{n}_e) = I(\mathbf{n}_c) + md^2$$
$$= \frac{1}{6}ma^2 + m\left(\frac{\sqrt{2}a}{2}\right)^2$$
$$= \frac{2}{3}ma^2.$$

Theorem 4. (Perpendicular Axis Theorem) *Suppose a rigid body is in the shape of a thin plate (planar shape). Let x, y, z be the orthogonal coordinate axes and x and y axes lie in the plane of the body. Then*

$$I_z = I_x + I_y.$$

Note this theorem applies only to planar rigid bodies. We can use the cube as a simple counterexample. By symmetry, we can easily infer that its moment of inertia along all three axes will be the same. Namely, $I_x = I_y = I_z$, and it is obvious that $I_z \neq I_x + I_y$. This is because the cube is not a planar body.

§5. Ellipsoid of a Tensor

Question. What is the moment of inertia of the cube with respect to the diagonal running through its center (Figure 8.2)?

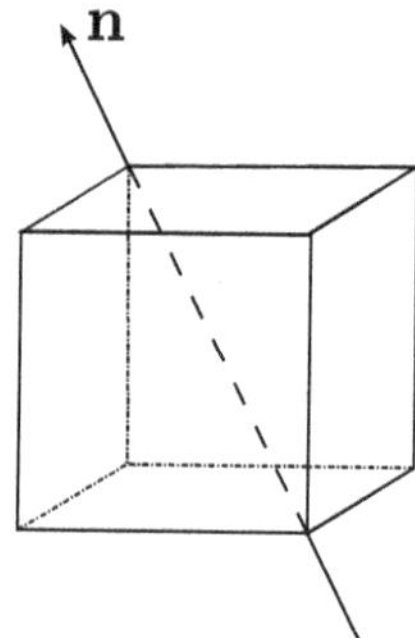

Figure 8.2   A cube and the diagonal axis

Of course, we can find this by brute-force integration according to Eq. 8.11. The calculation is more difficult because the boundary surfaces are complex with respect to the diagonal.

A more general method is to find the six components of the symmetric inertia matrix. Then given any axis represented by a unit vector $\mathbf{n}$, the scalar moment of inertia is

$$I(\mathbf{n}) = \mathbf{n} \cdot \mathscr{I} \cdot \mathbf{n}. \tag{8.12}$$

Suppose the unit vector $\mathbf{n}$ has components $\mathbf{n} = (\alpha, \beta, \gamma)$. Then the scalar moment of inertia relative to axis $\mathbf{n}$ is

$$I(\mathbf{n}) = I_{11}\alpha^2 + I_{22}\beta^2 + I_{33}\gamma^2 + 2I_{12}\alpha\beta + 2I_{23}\beta\gamma + 2I_{31}\gamma\alpha.$$

Suppose we have a desired condition,

$$T = \frac{1}{2}\mathscr{I}(\boldsymbol{\omega}) = \frac{1}{2},$$

or equivalently,

$$\mathscr{I}(\boldsymbol{\omega}) = \boldsymbol{\omega} \cdot \mathscr{I} \cdot \boldsymbol{\omega} = 1, \tag{8.13}$$

what condition must $\boldsymbol{\omega}$ meet?

If we write the angular velocity $\boldsymbol{\omega} = (\omega_x, \omega_y, \omega_z)$ in the component form, Eq. 8.13 becomes

$$I_{11}\omega_x^2 + I_{22}\omega_y^2 + I_{33}\omega_z^2 + 2I_{12}\omega_x\omega_y + 2I_{23}\omega_y\omega_z + 2I_{31}\omega_z\omega_x = 1. \tag{8.14}$$

The points $(\omega_x, \omega_y, \omega_z)$ satisfying the above equation in the space of $\boldsymbol{\omega}$ is an ellipsoid, which is called the **ellipsoid of inertia**.

Let ω be the magnitude and $\mathbf{n}$ the unit direction vector of $\boldsymbol{\omega}$. That is, $\boldsymbol{\omega} = \omega\mathbf{n}$. Then Eq. 8.13 becomes

$$\omega^2 \left(\mathbf{n} \cdot \mathcal{I} \cdot \mathbf{n}\right) = \omega^2 I(\mathbf{n}) = 1.$$

Therefore,

$$I(\mathbf{n}) = \frac{1}{\omega^2}. \tag{8.15}$$

This means that if the distance from the origin to the surface of this ellipsoid is ω, then the moment of inertia around this axis is $1/\omega^2$, according to Eq. 8.15.

The ellipsoid has three perpendicular principal axes, along the directions of three eigenvectors of $\mathcal{I}$. If we choose the coordinate axes x, y, z along these principal axes, the inertia matrix is in diagonal form and the ellipsoid of inertia takes the form

$$I_{11}\omega_x^2 + I_{22}\omega_y^2 + I_{33}\omega_z^2 = 1.$$

The scalar moment of inertia with respect to these principal axes are called **principal moments of inertia.**

Now let us answer the question in the beginning of this section. Because of symmetry, all three principal axes of the ellipsoid of inertia of the cube must be equal. Hence the ellipsoid of inertia for the cube is a sphere. This means its scalar moment of inertia relative to any axis through the center must be the same, which is $\frac{1}{6}ma^2$.

This method is not limited to the inertia tensor. It can be applied to any quadratic form and the ellipsoid is called the **ellipsoid of the quadratic form**, or **ellipsoid of the tensor**.

Chapter 9

Electrodynamics: The EM Field Tensor

The electric field and magnetic field are integral parts of the electromagnetic field. H. Minkowski [(1908)] unified the electric and magnetic field strengths into the electromagnetic field tensor, which was called a "vector of the second kind" (because it has 6 dimensions) by himself, or a 6-vector by A. Sommerfeld, to distinguish from a "vector of the first kind" (4 dimensions). Using the electromagnetic field tensor, the Maxwell equations can be written in a compact covariant form. Secs. 2 and 3 are based on a recent paper of the author [Guo (2021)]. The electromagnetic field tensor is an antisymmetric tensor, or a differential 2-form. Using the language of exterior calculus developed by É. Cartan, the Maxwell equations can be written in an even more compact form. Exterior calculus and differential forms are not in the scope of this book. Secs. 4 and 5 are just for the reader's reference, and also a motivation for further studies.

§1. Electrodynamics in Tensor Formulation

For simplicity, we discuss electromagnetic fields in vacuum only. The electromagnetic field is represented by the electric field strength $\mathbf{E}$ and the magnetic field strength $\mathbf{B}$ at every point in space.

Suppose in reference frame K, we adopt and rationalized natural units (Heaviside-Lorentz units and $c = 1$) and the coordinates $x^0 = t, x^1 = x, x^2 = y, x^3 = z$ (a signature of $+ - --$). The Maxwell equations in vacuum are in the form

$$\nabla \times \mathbf{B} - \partial_t \mathbf{E} = \mathbf{J},$$
$$\nabla \cdot \mathbf{E} = \rho,$$
$$\nabla \times \mathbf{E} + \partial_t \mathbf{B} = 0,$$
$$\nabla \cdot \mathbf{B} = 0. \tag{9.1}$$

When the sources are zero, $\mathbf{j} = 0$ and $\rho = 0$, the electromagnetic wave equations are

$$\left(\partial_t^2 - \nabla^2\right) \mathbf{E} = 0,$$
$$\left(\partial_t^2 - \nabla^2\right) \mathbf{B} = 0. \tag{9.2}$$

Suppose reference frame K$'$ moves at velocity v with respect to K in the x-direction. Under Lorentz transformation

$$t' = \gamma(t - vx),$$
$$x' = \gamma(x - vt),$$
$$y' = y,$$
$$z' = z, \tag{9.3}$$

where $\gamma = \sqrt{1 - v^2}$, together with field transformation

$$
\begin{aligned}
E_1' &= E_1, & B_1' &= B_1, \\
E_2' &= \gamma(E_2 - vB_3), & B_2' &= \gamma(B_2 + vE_3), \\
E_3' &= \gamma(E_3 + vB_2), & B_3' &= \gamma(B_3 - vE_2),
\end{aligned}
\tag{9.4}
$$

the Maxwell equations in frame K$'$ in terms of field strengths $\mathbf{E}'$ and $\mathbf{B}'$ satisfy the same equations as in Eq. 9.1. The field transformation Eq. 9.4 can be written in 3-vector form:

$$\mathbf{E}' = \gamma(\mathbf{E} + \mathbf{v} \times \mathbf{B}) - (\gamma - 1)\hat{\mathbf{v}}(\hat{\mathbf{v}} \cdot \mathbf{E}),$$
$$\mathbf{B}' = \gamma(\mathbf{B} - \mathbf{v} \times \mathbf{E}) - (\gamma - 1)\hat{\mathbf{v}}(\hat{\mathbf{v}} \cdot \mathbf{B}), \tag{9.5}$$

where $\hat{\mathbf{v}}$ is the unit vector in the direction of $\mathbf{v}$.

The tensor formulation of Maxwell equations is due to Minkowski [(1908)]. Using the electromagnetic tensor defined in the following,

$$F^{\mu\nu} = \begin{bmatrix} 0 & -E_1 & -E_2 & -E_3 \\ E_1 & 0 & -B_3 & B_2 \\ E_2 & B_3 & 0 & -B_1 \\ E_3 & -B_2 & B_1 & 0 \end{bmatrix}, \tag{9.6}$$

$$F_{\mu\nu} = \begin{bmatrix} 0 & E_1 & E_2 & E_3 \\ -E_1 & 0 & -B_3 & B_2 \\ -E_2 & B_3 & 0 & -B_1 \\ -E_3 & -B_2 & B_1 & 0 \end{bmatrix}, \tag{9.7}$$

Maxwell equations can be written in a compact and covariant form,

$$\sum_{\mu} \partial_\mu F^{\nu\mu} = J^\nu, \tag{9.8}$$

$$\partial_\lambda F_{\mu\nu} + \partial_\nu F_{\lambda\mu} + \partial_\mu F_{\nu\lambda} = 0. \tag{9.9}$$

Note Eq. 9.8 translates to four scalar equations. Eq. 9.9 might look like $4 \times 4 \times 4 = 64$ scalar equations at the first glance, but they are actually four independent scalar equations.

§2. Electrodynamics under Galilean Transformation

* **Misconception:** Galilean transformation is incompatible with the Maxwell equations.

The fact is that the Maxwell equations are not "invariant" under the Galilean transformation. "Non-invariant" and "incompatible" are different concepts. The latter means "logically contradicting", while the former does not. Furthermore, when we say that the Maxwell equations are not "invariant" under the Galilean transformation, we mean Maxwell equations in the 3-vector form are not invariant. When they are written in the tensor form in 4-dimensional spacetime as in Eqs. 9.8 and 9.9, the form is covariant under any linear transformations, including Lorentz transformation and Galilean transformation. The only difference is that Lorentz transformation is pseudo-orthogonal while the Galilean transformation is not. Both are valid and the difference is only a matter of convenience (see more about the equivalence of Galilean transformation and Lorentz transformation in Sec. 6 of Chap. 10).

The following is based on a recent paper.[1] See more details in [Guo (2021)]. We shall give a formulation of Maxwell equations, as well as the equations of electromagnetic waves in vacuum under Galilean transformation

$$
\begin{aligned}
t' &= t, \\
x' &= x - vt, \\
y' &= y, \\
z' &= z.
\end{aligned}
\tag{9.10}
$$

The Galilean transformation can be written as

$$
(x')^{\mu} = \sum_{\nu} \Lambda^{\mu}{}_{\nu} x^{\nu},
\tag{9.11}
$$

where Λ is a matrix

$$
\Lambda =
\begin{bmatrix}
1 & 0 & 0 & 0 \\
-v & 1 & 0 & 0 \\
0 & 0 & 1 & 0 \\
0 & 0 & 0 & 1
\end{bmatrix}.
\tag{9.12}
$$

In reference frame K, the spacetime quadratic form is

$$
\begin{aligned}
ds^2 &= \sum_{\mu\nu} g_{\mu\nu} dx^{\mu} dx^{\nu} \\
&= dt^2 - dx^2 - dy^2 - dz^2.
\end{aligned}
\tag{9.13}
$$

In reference frame K',

$$
\begin{aligned}
ds^2 &= \sum_{\mu\nu} g'_{\mu\nu} (dx')^{\mu} (dx')^{\nu} \\
&= \left(1 - v^2\right) (dt')^2 - 2v\,dt'dx' - (dx')^2 - (dy')^2 - (dz')^2,
\end{aligned}
\tag{9.14}
$$

with the pseudo-metric tensor

$$
g'_{\mu\nu} =
\begin{bmatrix}
1/\gamma^2 & -v & 0 & 0 \\
-v & -1 & 0 & 0 \\
0 & 0 & -1 & 0 \\
0 & 0 & 0 & -1
\end{bmatrix}.
\tag{9.15}
$$

---

[1]Guo, H. (2021). A New Paradox and the Reconciliation of Lorentz and Galilean Transformations, *Synthese*, https://doi.org/10.1007/s11229-021-03155-y (open access).

Under Galilean transformation, the differential operators transform according to

$$\partial_{t'} = \partial_t + \mathbf{v} \cdot \nabla,$$
$$\nabla' = \nabla. \tag{9.16}$$

The charge and current transform according to

$$\rho' = \rho,$$
$$\mathbf{j}' = \mathbf{j} - \rho\mathbf{v}. \tag{9.17}$$

In reference frame K$'$ under Galilean transformation Eq. 9.10, the contravariant field tensor is

$$(F')^{\mu\nu} = \sum_{\alpha\beta} \Lambda^{\mu}{}_{\alpha} \Lambda^{\nu}{}_{\beta} F^{\alpha\beta}$$

$$= \begin{bmatrix} 0 & -E_1 & -E_2 & -E_3 \\ E_1 & 0 & -(B_3 - vE_2) & B_2 + vE_3 \\ E_2 & B_3 - vE_2 & 0 & -B_1 \\ E_3 & -(B_2 + vE_3) & B_1 & 0 \end{bmatrix}. \tag{9.18}$$

The covariant field tensor is

$$F'_{\mu\nu} = \sum_{\alpha\beta} g'_{\mu\alpha} g'_{\nu\beta} F'^{\alpha\beta}$$

$$= \begin{bmatrix} 0 & E_1 & E_2 - vB_3 & E_3 + vB_2 \\ -E_1 & 0 & -B_3 & B_2 \\ -(E_2 - vB_3) & B_3 & 0 & -B_1 \\ -(E_3 + vB_2) & -B_2 & B_1 & 0 \end{bmatrix}. \tag{9.19}$$

Note Eq. 9.10 is only the transformation of space and time. To find the Maxwell equations in reference frame K$'$, we also need to figure out how the electromagnetic field transforms. The field transformation Eq. 9.4 that works together with Lorentz transformation is motivated to have the form of Maxwell equations invariant. In the case of Galilean transformation, we know we cannot keep the form of Maxwell equations invariant. So this clue is lost. In my opinion, the field transformation is not a law of nature. It can be arbitrary by convention. Our consideration is again convenience rather than absolute truth. The EM field as a whole is described by the field tensor $F^{\mu\nu}$. We view this field tensor $F^{\mu\nu}$ more essential than the 3-vectors $\mathbf{E}$ and $\mathbf{B}$, which are just some names of the components of $F^{\mu\nu}$. The transformation of the field is not directly observable. What matters to the physical phenomena is the field together with how the field interacts with matter (Lorentz force law). In the following we shall explore different field transformation laws.

2.1   EM Field in the Form of Contravariant Tensor $F^{\mu\nu}$

We choose to define the electric and magnetic fields in reference frame K$'$ as

$$\begin{bmatrix} E_1' \\ E_2' \\ E_3' \end{bmatrix} \overset{\text{def}}{=} \begin{bmatrix} (F')^{10} \\ (F')^{20} \\ (F')^{30} \end{bmatrix}, \quad \begin{bmatrix} B_1' \\ B_2' \\ B_3' \end{bmatrix} \overset{\text{def}}{=} \begin{bmatrix} (F')^{32} \\ (F')^{13} \\ (F')^{21} \end{bmatrix}. \tag{9.20}$$

This means

$$(F')^{\mu\nu} = \begin{bmatrix} 0 & -E_1' & -E_2' & -E_3' \\ E_1' & 0 & -B_3' & B_2' \\ E_2' & B_3' & 0 & -B_1' \\ E_3' & -B_2' & B_1' & 0 \end{bmatrix}. \tag{9.21}$$

Comparing with Eq. 9.18, we find the field transformation in 3-vector form is

$$\mathbf{E}' = \mathbf{E},$$
$$\mathbf{B}' = \mathbf{B} - \mathbf{v} \times \mathbf{E}. \tag{9.22}$$

The inverse transformation of Eq. 9.22 is

$$\mathbf{E} = \mathbf{E}',$$
$$\mathbf{B} = \mathbf{B}' + \mathbf{v} \times \mathbf{E}'. \tag{9.23}$$

The Maxwell equations in 3-vector form are

$$\nabla' \times (\mathbf{B}' + \mathbf{v} \times \mathbf{E}') - (\partial_{t'} - \mathbf{v} \cdot \nabla') \mathbf{E}' = \mathbf{j}' + \rho' \mathbf{v},$$
$$\nabla' \cdot \mathbf{E}' = \rho',$$
$$\nabla' \times \mathbf{E}' + (\partial_{t'} - \mathbf{v} \cdot \nabla') (\mathbf{B}' + \mathbf{v} \times \mathbf{E}') = 0,$$
$$\nabla' \cdot (\mathbf{B}' + \mathbf{v} \times \mathbf{E}') = 0. \tag{9.24}$$

This can be simplified to

$$\nabla' \times \mathbf{B}' - \partial_{t'} \mathbf{E}' = \mathbf{j}',$$
$$\nabla \cdot \mathbf{E}' = \rho',$$
$$\nabla' \times \mathbf{E}' + \partial_{t'} \mathbf{B}' = (\mathbf{v} \cdot \nabla') (\mathbf{B}' + \mathbf{v} \times \mathbf{E}') - \mathbf{v} \times \partial_{t'} \mathbf{E}',$$
$$\nabla \cdot \mathbf{B}' = -\nabla' \cdot (\mathbf{v} \times \mathbf{E}'). \tag{9.25}$$

The first two equations are Galilean invariant. When the sources are zero, the equations of electromagnetic wave in vacuum are

$$\left[(\partial_{t'} - \mathbf{v} \cdot \nabla')^2 - \nabla'^2 \right] \mathbf{E}' = 0,$$
$$\left[(\partial_{t'} - \mathbf{v} \cdot \nabla')^2 - \nabla'^2 \right] (\mathbf{B}' + \mathbf{v} \times \mathbf{E}') = 0. \tag{9.26}$$

2.2   EM Field in the Form of Covariant Tensor $F_{\mu\nu}$

We choose to define the electric and magnetic fields in reference frame K$'$ as

$$\begin{bmatrix} E'_1 \\ E'_2 \\ E'_3 \end{bmatrix} \stackrel{\text{def}}{=} \begin{bmatrix} F'_{01} \\ F'_{02} \\ F'_{03} \end{bmatrix}, \quad \begin{bmatrix} B'_1 \\ B'_2 \\ B'_3 \end{bmatrix} \stackrel{\text{def}}{=} \begin{bmatrix} F'_{32} \\ F'_{13} \\ F'_{21} \end{bmatrix}. \tag{9.27}$$

We also define auxiliary dual fields $\mathscr{E}'_1, \mathscr{E}'_2, \mathscr{E}'_3, \mathscr{B}'_1, \mathscr{B}'_2, \mathscr{B}'_3$ such that

$$(F')^{\mu\nu} \stackrel{\text{def}}{=} \begin{bmatrix} 0 & -\mathscr{E}'_1 & -\mathscr{E}'_2 & -\mathscr{E}'_3 \\ \mathscr{E}'_1 & 0 & -\mathscr{B}'_3 & \mathscr{B}'_2 \\ \mathscr{E}'_2 & \mathscr{B}'_3 & 0 & -\mathscr{B}'_1 \\ \mathscr{E}'_3 & -\mathscr{B}'_2 & \mathscr{B}'_1 & 0 \end{bmatrix}, \quad F'_{\mu\nu} = \begin{bmatrix} 0 & E'_1 & E'_2 & E'_3 \\ -E'_1 & 0 & -B'_3 & B'_2 \\ -E'_2 & B'_3 & 0 & -B'_1 \\ -E'_3 & -B'_2 & B'_1 & 0 \end{bmatrix}. \tag{9.28}$$

Comparing with Eq. 9.19, we find the field transformation in 3-vector form is

$$\mathbf{E}' = \mathbf{E} + \mathbf{v} \times \mathbf{B}, \tag{9.29}$$
$$\mathbf{B}' = \mathbf{B}.$$

The dual field vectors $\mathscr{E}'$ and $\mathscr{B}'$ are related to $\mathbf{E}'$ and $\mathbf{B}'$ through

$$\mathscr{E}' = \mathbf{E}' - \mathbf{v} \times \mathbf{B}', \tag{9.30}$$
$$\mathscr{B}' = \mathbf{B}' - \mathbf{v} \times \left(\mathbf{E}' - \mathbf{v} \times \mathbf{B}'\right).$$

The inverse transformation of Eq. 9.29 is

$$\mathbf{E} = \mathbf{E}' - \mathbf{v} \times \mathbf{B}', \tag{9.31}$$
$$\mathbf{B} = \mathbf{B}'.$$

The Maxwell equations in 3-vector form are

$$\nabla' \times \mathbf{B}' - \left(\partial_{t'} - \mathbf{v} \cdot \nabla'\right)\left(\mathbf{E}' - \mathbf{v} \times \mathbf{B}'\right) = \mathbf{j}' + \rho'\mathbf{v},$$
$$\nabla' \cdot \left(\mathbf{E}' - \mathbf{v} \times \mathbf{B}'\right) = \rho',$$
$$\nabla' \times \left(\mathbf{E}' - \mathbf{v} \times \mathbf{B}'\right) + \left(\partial_{t'} - \mathbf{v} \cdot \nabla'\right)\mathbf{B}' = 0, \tag{9.32}$$
$$\nabla' \cdot \mathbf{B}' = 0.$$

This can be simplified to

$$\nabla' \times \mathbf{B}' - \partial_{t'}\mathbf{E}' = \mathbf{j}' + \nabla' \times \left[\mathbf{v} \times \left(\mathbf{E}' - \mathbf{v} \times \mathbf{B}'\right)\right] + \mathbf{v} \times \left(\nabla' \times \mathbf{E}'\right),$$
$$\nabla' \cdot \mathbf{E}' = \rho' + \nabla' \cdot \left(\mathbf{v} \times \mathbf{B}'\right),$$
$$\nabla' \times \mathbf{E}' + \partial_{t'}\mathbf{B}' = 0, \tag{9.33}$$
$$\nabla' \cdot \mathbf{B}' = 0.$$

The last two equations are Galilean invariant. When the sources are zero, the equations of electromagnetic wave in vacuum are

$$\left[\left(\partial_{t'} - \mathbf{v} \cdot \nabla'\right)^2 - \nabla'^2\right]\left(\mathbf{E}' - \mathbf{v} \times \mathbf{B}'\right) = 0,$$
$$\left[\left(\partial_{t'} - \mathbf{v} \cdot \nabla'\right)^2 - \nabla'^2\right]\mathbf{B}' = 0. \tag{9.34}$$

2.3   EM Field in the Form of a Mixture of $F^{\mu\nu}$ and $F_{\mu\nu}$

We choose to define the electric and magnetic fields in reference frame K' as

$$\begin{bmatrix} E_1' \\ E_2' \\ E_3' \end{bmatrix} \overset{\text{def}}{=} \begin{bmatrix} (F')^{10} \\ (F')^{20} \\ (F')^{30} \end{bmatrix}, \quad \begin{bmatrix} B_1' \\ B_2' \\ B_3' \end{bmatrix} \overset{\text{def}}{=} \begin{bmatrix} F_{32}' \\ F_{13}' \\ F_{21}' \end{bmatrix}. \tag{9.35}$$

We also define auxiliary dual fields $\mathscr{E}_1', \mathscr{E}_2', \mathscr{E}_3', \mathscr{B}_1', \mathscr{B}_2', \mathscr{B}_3'$ such that

$$(F')^{\mu\nu} = \begin{bmatrix} 0 & -E_1' & -E_2' & -E_3' \\ E_1' & 0 & -\mathscr{B}_3' & \mathscr{B}_2' \\ E_2' & \mathscr{B}_3' & 0 & -\mathscr{B}_1' \\ E_3' & -\mathscr{B}_2' & \mathscr{B}_1' & 0 \end{bmatrix}, \quad F_{\mu\nu}' = \begin{bmatrix} 0 & \mathscr{E}_1' & \mathscr{E}_2' & \mathscr{E}_3' \\ -\mathscr{E}_1' & 0 & -B_3' & B_2' \\ -\mathscr{E}_2' & B_3' & 0 & -B_1' \\ -\mathscr{E}_3' & -B_2' & B_1' & 0 \end{bmatrix}. \tag{9.36}$$

Comparing with Eqs. 9.18 and 9.19, we find the field transformation in 3-vector form is

$$\begin{aligned} \mathbf{E}' &= \mathbf{E}, \\ \mathbf{B}' &= \mathbf{B}. \end{aligned} \tag{9.37}$$

We find the dual field vectors $\mathscr{E}'$ and $\mathscr{B}'$ are related to $\mathbf{E}'$ and $\mathbf{B}'$ through

$$\begin{aligned} \mathscr{E}' &= \mathbf{E}' + \mathbf{v} \times \mathbf{B}', \\ \mathscr{B}' &= \mathbf{B}' - \mathbf{v} \times \mathbf{E}'. \end{aligned} \tag{9.38}$$

The Maxwell equations in 3-vector form are

$$\begin{aligned} \nabla' \times \mathbf{B}' - \left(\partial_{t'} - \mathbf{v} \cdot \nabla' \right) \mathbf{E}' &= \mathbf{j}' + \rho' \mathbf{v}, \\ \nabla' \cdot \mathbf{E}' &= \rho', \\ \nabla' \times \mathbf{E}' + \left(\partial_{t'} - \mathbf{v} \cdot \nabla' \right) \mathbf{B}' &= 0, \\ \nabla' \cdot \mathbf{B}' &= 0. \end{aligned} \tag{9.39}$$

This can be simplified to

$$\begin{aligned} \nabla' \times \mathbf{B}' - \partial_{t'} \mathbf{E}' &= \mathbf{j}' - \left(\mathbf{v} \cdot \nabla' \right) \mathbf{E}', \\ \nabla' \cdot \mathbf{E} &= \rho', \\ \nabla' \times \mathbf{E}' + \partial_{t'} \mathbf{B}' &= \left(\mathbf{v} \cdot \nabla' \right) \mathbf{B}', \\ \nabla \cdot \mathbf{B}' &= 0. \end{aligned} \tag{9.40}$$

The second and the fourth equations are Galilean invariant. When the sources are zero, the equations of electromagnetic wave in vacuum are

$$\begin{aligned} \left[\left(\partial_{t'} - \mathbf{v} \cdot \nabla' \right)^2 - \nabla'^2 \right] \mathbf{E}' &= 0, \\ \left[\left(\partial_{t'} - \mathbf{v} \cdot \nabla' \right)^2 - \nabla'^2 \right] \mathbf{B}' &= 0. \end{aligned} \tag{9.41}$$

If $\mathbf{v}$ is in the direction of x, Eq. 9.41 can be simplified to

$$\frac{1}{\gamma^2}\partial_{x'}^2\mathbf{E}' + \partial_{y'}^2\mathbf{E}' + \partial_{z'}^2\mathbf{E}' + 2v\partial_{t'}\partial_{x'}\mathbf{E}' - \partial_{t'}^2\mathbf{E}' = 0,$$

$$\frac{1}{\gamma^2}\partial_{x'}^2\mathbf{B}' + \partial_{y'}^2\mathbf{B}' + \partial_{z'}^2\mathbf{B}' + 2v\partial_{t'}\partial_{x'}\mathbf{B}' - \partial_{t'}^2\mathbf{B}' = 0. \tag{9.42}$$

2.4 EM Field in the Form of a Mixture of $F^{\mu\nu}$ and $F_{\mu\nu}$

We choose to define the electric and magnetic fields in reference frame K' as

$$\begin{bmatrix} E_1' \\ E_2' \\ E_3' \end{bmatrix} \overset{\text{def}}{=} \begin{bmatrix} F_{01}' \\ F_{02}' \\ F_{03}' \end{bmatrix}, \quad \begin{bmatrix} B_1' \\ B_2' \\ B_3' \end{bmatrix} \overset{\text{def}}{=} \begin{bmatrix} (F')^{32} \\ (F')^{13} \\ (F')^{21} \end{bmatrix}. \tag{9.43}$$

We also define auxiliary dual fields $\mathcal{E}_1', \mathcal{E}_2', \mathcal{E}_3', \mathcal{B}_1', \mathcal{B}_2', \mathcal{B}_3'$ such that

$$(F')^{\mu\nu} \overset{\text{def}}{=} \begin{bmatrix} 0 & -\mathcal{E}_1' & -\mathcal{E}_2' & -\mathcal{E}_3' \\ \mathcal{E}_1' & 0 & -B_3' & B_2' \\ \mathcal{E}_2' & B_3' & 0 & -B_1' \\ \mathcal{E}_3' & -B_2' & B_1' & 0 \end{bmatrix}, \quad F_{\mu\nu}' \overset{\text{def}}{=} \begin{bmatrix} 0 & E_1' & E_2' & E_3' \\ -E_1' & 0 & \mathcal{B}_3' & \mathcal{B}_2' \\ -E_2' & \mathcal{B}_3' & 0 & -\mathcal{B}_1' \\ -E_3' & -\mathcal{B}_2' & \mathcal{B}_1' & 0 \end{bmatrix}. \tag{9.44}$$

Comparing with Eqs. 9.19 and 9.18, we find the field transformation in 3-vector form is

$$\mathbf{E}' = \mathbf{E} + \mathbf{v} \times \mathbf{B},$$

$$\mathbf{B}' = \mathbf{B} - \mathbf{v} \times \mathbf{E}. \tag{9.45}$$

The dual field vectors $\mathcal{E}'$ and $\mathcal{B}'$ are related to $\mathbf{E}$ and $\mathbf{B}$ through

$$\mathcal{E}' = \mathbf{E},$$

$$\mathcal{B}' = \mathbf{B}. \tag{9.46}$$

To solve the Maxwell equations in reference frame K', it is easier to write them in terms of $\mathbf{E}$ and $\mathbf{B}$ but in coordinates (x', y', z', t'),

$$\nabla' \times \mathbf{B} - (\partial_{t'} - \mathbf{v} \cdot \nabla')\mathbf{E} = \mathbf{j}' + \rho'\mathbf{v},$$

$$\nabla' \cdot \mathbf{E} = \rho',$$

$$\nabla' \times \mathbf{E} + (\partial_{t'} - \mathbf{v} \cdot \nabla')\mathbf{B} = 0,$$

$$\nabla' \cdot \mathbf{B} = 0. \tag{9.47}$$

This is basically the same as Eq. 9.39. After solving for $\mathbf{E}$ and $\mathbf{B}$ in coordinates (x', y', z', t'), we can obtain $\mathbf{E}'$ and $\mathbf{B}'$ using field transformation Eq. 9.45.

§3. Electrodynamics in Rotating Reference Frames

Electrodynamics in rotating reference frames is rarely addressed in textbooks. In a rotating reference frame, we can use the transformation in polar coordinates

$$
\begin{aligned}
t' &= t, \\
r' &= r, \\
\varphi' &= \varphi - \omega t, \\
z' &= z.
\end{aligned}
\tag{9.48}
$$

We may call this the rotational Galilean transformation, or Galilean-like transformation. We may call any transformation with the characteristic $t' = t$ a Galilean-like transformation. Under transformation Eq. 9.48, the differential operators transform as

$$
\begin{aligned}
\partial_{t'} &= \partial_t + \mathbf{v} \cdot \nabla, \quad \text{where } \mathbf{v} = \boldsymbol{\omega} \times \mathbf{r}, \\
\nabla' &= \nabla.
\end{aligned}
\tag{9.49}
$$

The charge and current transform as

$$
\begin{aligned}
\rho' &= \rho, \\
\mathbf{j}' &= \mathbf{j} - \rho \mathbf{v}, \qquad \text{where } \mathbf{v} = \boldsymbol{\omega} \times \mathbf{r}.
\end{aligned}
\tag{9.50}
$$

These have the same apparent form as Eqs. 9.16 and 9.17 for the Galilean transformation, except now $\mathbf{v} = \boldsymbol{\omega} \times \mathbf{r}$ is a variable, rather than a constant. For all the discussions of electrodynamics in the previous section, if we replace the constant $\mathbf{v}$ with $\mathbf{v} = \boldsymbol{\omega} \times \mathbf{r}$, we can obtain the electrodynamic equations for the rotating reference frame.

Among inertial reference frames, we do have a choice of Lorentz transformation and Galilean transformation. Both are valid but the Galilean transformation is not convenient. For rotating reference frames, we could choose any transformation as well (but neither Lorentz transformation nor Galilean transformation is for the rotating reference frame). For a rotating reference frame, no transformation seems convenient. This is due to the intrinsic complexity of rotating frames. However, among all the inconvenient transformations for the rotating reference frames, the rotational Galilean transformation Eq. 9.48 could be the least inconvenient.

*§4. Maxwell Equations in Exterior Differential Forms

The electromagnetic field tensor in Eq. 9.6 is a special type of tensor. It is easy to see that it is antisymmetric,

$$F_{\mu\nu} = -F_{\nu\mu}.$$

In fact, it is a differential 2-form. Maxwell equations can be written in even more compact form using differential forms. Exterior calculus and differential forms are not in the scope of this book, but we would like to show it here for the reader's reference, and for a motivation for further studies.

Let U be the 4-potential. The electromagnetic field tensor is defined to be

$$F^{\mu\nu} = \partial^\nu U^\mu - \partial^\mu U^\nu.$$

F is in a 6-dimensional linear subspace of the 16-dimensional tensor space. A. Sommerfeld called it a 6-vector.

In differential forms,

$$F = dU.$$

The potential U is a differential 1-form and F is a differential 2-form with two parts, the electric field E and the magnetic field B,

$$F = E + B,$$

where

$$E = E_x dx \wedge dt + E_y dy \wedge dt + E_z dz \wedge dt,$$
$$B = B_x dy \wedge dz + B_y dz \wedge dx + B_z dx \wedge dy.$$

Theorem 1. *Maxwell equations can be written as follows using exterior derivative and differential forms:*

$$d(*F) = *J, \tag{9.51}$$
$$dF = 0, \tag{9.52}$$

where d is the exterior derivative operator, $*$ is the Hodge dual operator and J is the current 1-form

$$J = J_x dx + J_y dy + J_z dz + J_t dt.$$

Note in coordinate form,

$$d(\cdot) \stackrel{\text{def}}{=} \frac{\partial(\cdot)}{\partial t} dt + \frac{\partial(\cdot)}{\partial x} dx + \frac{\partial(\cdot)}{\partial y} dy + \frac{\partial(\cdot)}{\partial z} dz,$$

where the dot is the place holder for a scalar field.

Theorem 2. (Conservation of charge or continuity equation)

$$d(*J) = 0.$$

A simple proof of Theorem 2 is to take the exterior derivative of both sides of Eq. 9.51,

$$d(*J) = d^2(*F) = 0,$$

because

$$d^2 \equiv 0.$$

*§5. Proposal of New Notation $d\wedge$ for Exterior Derivative

It is the standard notation in current literature to use $d\omega^k$ to denote the exterior derivative of differential k-form ω^k. I propose a new mnemonic notation

$$d \wedge \omega^k,$$

in lieu of $d\omega^k$, where

$$d \stackrel{\text{def}}{=} \frac{\partial}{\partial x_1} dx_1 + \ldots + \frac{\partial}{\partial x_n} dx_n$$

is a "symbolic 1-form operator". Well, an ordinary 1-form ω^1 itself (in the cotangent space at a fixed point of the differentiable manifold M) is an operator, operating on tangent vectors: $\omega^1 : T_p(M) \to \mathbb{R}$. In coordinate form,

$$\omega^1 = a_1 dx_1 + \ldots + a_n dx_n,$$

where $a_1, \ldots, a_n$ are constants. By a "symbolic 1-form operator", I mean it can be treated symbolically as a 1-form expressed as the linear combination of local basis vectors $dx_1, \ldots, dx_n$, but the coefficients are differential operators $\dfrac{\partial}{\partial x_1}, \ldots, \dfrac{\partial}{\partial x_n}$ operating on scalar fields on the manifold, instead of constants $a_1, \ldots, a_n$.

Let us look at an analogy with the nabla symbolic vector operator introduced by Hamilton,

$$\nabla \overset{\text{def}}{=} \frac{\partial}{\partial x}\mathbf{i} + \frac{\partial}{\partial y}\mathbf{j} + \frac{\partial}{\partial z}\mathbf{k}.$$

Traditionally, three differential operators are used: $\text{grad}f$, $\text{div}\mathbf{A}$ and $\text{curl}\mathbf{A}$, for gradient of a scalar field f, and the divergence and the curl of a vector field $\mathbf{A}$. These three operations can be denoted with a single nabla vector operator ∇, together with vector dot product and cross product,

$$\nabla f \overset{\text{def}}{=} \text{grad}f,$$

$$\nabla \cdot \mathbf{A} \overset{\text{def}}{=} \text{div}\mathbf{A},$$

$$\nabla \times \mathbf{A} \overset{\text{def}}{=} \text{curl}\mathbf{A},$$

to express these three operations of gradient, divergence and curl, as if ∇ is a vector, and it can have a symbolic dot product and cross product operation with a vector field.

It is a theorem in vector analysis that

$$\text{curl}\,(\text{grad}f) = 0,$$

$$\text{div}\,(\text{curl}\mathbf{A}) = 0,$$

for any scalar field f and vector field $\mathbf{A}$. We can see the advantage of the nabla notation when this theorem is expressed using the nabla operator:

$$\nabla \times \nabla f = 0,$$

$$\nabla \cdot (\nabla \times \mathbf{A}) = 0.$$

These are analogous to the vector identities

$$\mathbf{B} \times \mathbf{B} = 0,$$

$$\mathbf{B} \cdot (\mathbf{B} \times \mathbf{A}) = 0,$$

for ordinary vectors $\mathbf{B}$ and $\mathbf{A}$.

The $\wedge$ in $d \wedge \omega^k$ means the "formal wedge product". Even if d is a differential operator, it is treated symbolically as a 1-form, hence it is apparent that

$$d \wedge \omega^k = \left(\frac{\partial}{\partial x_1}dx_1 + \ldots + \frac{\partial}{\partial x_n}dx_n \right) \wedge \omega^k$$

$$= \frac{\partial}{\partial x_1}dx_1 \wedge \omega^k + \ldots + \frac{\partial}{\partial x_n}dx_n \wedge \omega^k$$

is a $(k+1)$-form, because it is the wedge product of a 1-form with a k-form.

Compare with the nabla operator ∇ again. O. Heaviside called the nabla operator "the fictitious vector". Here in analogy, the exterior derivative operator d can be viewed as the "fictitious 1-form". Hence we can perform a "fictitious" or "formal" wedge product $d \wedge \omega^k$ of d with another differential form ω^k.

If f is a scalar field, then

$$d \wedge f = df = \frac{\partial f}{\partial x_1} dx_1 + \ldots + \frac{\partial f}{\partial x_n} dx_n.$$

At each point $p \in M$, $f \in \mathbb{R}$ is a scalar. This is in analogy to

$$\omega^1 \wedge f = f \wedge \omega^1 = f\omega^1 = a_1 f dx_1 + \ldots + a_n f dx_n,$$

where ω^1 is any ordinary 1-form and $a_1, \ldots, a_n$ are ordinary scalar constants.

If ω^k is any differential k-form written in the component form

$$\omega^k = \sum_{i_1,\ldots,i_k} f_{i_1 \cdots i_k}(x_1, \ldots, x_n) dx_{i_1} \wedge \ldots \wedge dx_{i_k},$$

then its exterior derivative is a $(k+1)$-form,

$$d \wedge \omega^k = \sum_{i_1,\ldots,i_k} \left(\frac{\partial f_{i_1 \cdots i_k}}{\partial x_1} dx_1 + \ldots + \frac{\partial f_{i_1 \cdots i_k}}{\partial x_n} dx_n \right) \wedge dx_{i_1} \wedge \ldots \wedge dx_{i_k}$$

$$= \sum_{i_1,\ldots,i_k} \left(\frac{\partial f_{i_1 \cdots i_k}}{\partial x_1} dx_1 \wedge dx_{i_1} \wedge \ldots \wedge dx_{i_k} \right.$$

$$\left. + \ldots + \frac{\partial f_{i_1 \cdots i_k}}{\partial x_n} dx_n \wedge dx_{i_1} \wedge \ldots \wedge dx_{i_k} \right).$$

It is a well known property that every exact form is a closed form. In the usual notation of d, it is expressed as

$$d^2 \equiv 0,$$

namely, $d\left(d\omega^k\right) = d^2 \omega^k \equiv 0$, for all differential forms ω^k.

We can see another advantage of this new notation $d\wedge$. Using this new notation, it becomes the following.

Theorem 3. *Every exact form $(d \wedge \omega^k)$ is a closed form. Namely*

$$d \wedge d \equiv 0,$$

or, $d \wedge d \wedge \omega^k \equiv 0$, for all differential forms ω^k.

It looks more intuitive using this new notation, because this is in analogy to

$$\omega^1 \wedge \omega^1 \equiv 0,$$

where ω^1 is an ordinary 1-form.

Theorem 4. *Maxwell equations can be written as follows using the new notation:*

$$d \wedge (*F) = *J,$$
$$d \wedge F = 0.$$

Theorem 5. (Conservation of charge) *The continuity equation can be written as follows using the new notation:*

$$d \wedge (*J) = 0.$$

A simple proof is that

$$d \wedge (*J) = d \wedge d \wedge (*F) = 0,$$

because

$$d \wedge d \equiv 0.$$

Chapter 10

Riemannian Geometry and
General Relativity

One geometry cannot be more true than another; it can only be more convenient.

— Henri Poincaré

One time standard cannot be more true than another; it can only be more convenient.

   Time is a human convention. There is no true time, or God-given time, or by whatever other names—natural time, physical time, cosmic time, etc. The phrase "physical time" is an oxymoron. God created matter. Man created time. Gravity does not curve spacetime. Man curves it.

— Hongyu Guo

Mathematics is the art of giving the same name to different things. Poetry is the art of giving different names to the same thing.

— Henri Poincaré

119

We conclude the book with an outlook on Riemannian geometry and general relativity in this chapter. We would like to clarify that the metric tensor is essential in Riemannian geometry not because it is a tensor, but rather because it is the key mathematical structure of a Riemannian manifold, which provides the notion of intrinsic distance (see Appendix 2). We can go by without calling it a tensor. It is just an inner product in the tangent space, or intuitively in an infinitesimal neighborhood of the manifold. The intrinsic view is very important to appreciate Riemannian geometry, which is a generalization of the intrinsic geometry of surfaces due to Gauss. Some subtle differences between the intrinsic view and extrinsic view, between differentiable manifolds and Riemannian manifolds, and between pseudo-Riemannian manifolds and Riemannian manifolds are emphasized. Sec. 7 is based on a recent paper [Guo (2021)],[1] which reflects a conventionalist view of the author on time and relativity. For further reading, the reader is referred to [Bishop and Goldberg (1980)] and [Guo (2014)].

§1. What Is "Curved Space" Exactly?

We often hear people say: "The curved space is beyond my imagination." If you are confused, it is not your fault. The term "curved space" is informal, and the term by itself is mystical and misleading. The formal term is Riemannian manifold.

Remark. "Curved space" is the generalization of curved surfaces. A 2-dimensional curved surface can be viewed as a subset of the 3-dimensional Euclidean space. Similarly, a 3-dimensional curved space can be viewed as a subset of the 4-dimensional Euclidean space. That is, the curved space curves in the 4-dimensional Euclidean space. However, this is the extrinsic view, meaning we look at this curved space from a higher dimensional Euclidean space. So curved space from the extrinsic view is not hard at all. If we apply this view to our real world, we find this 4-dimensional ambient Euclidean space does not exist, or there is no way to detect and measure it. The only way is to describe "the curved space" from within. This is the intrinsic view, which is harder. From the perspective of intrinsic geometry, "curved" just means "different", and nothing more. A "curved

---

[1]Guo, H. (2021). A New Paradox and the Reconciliation of Lorentz and Galilean Transformations, *Synthese*, https://doi.org/10.1007/s11229-021-03155-y (open access).

space" is just a "different space". Different from what? Different from the 3-dimensional Euclidean space. All we can see (or feel) is that its laws of geometry are different from those of Euclidean geometry. For example, the circumference of a circle is different from $2\pi r$. The Pythagorean theorem is no longer true. The curvature of space is nothing but a quantitative measure of the differences. The space may have a positive or negative curvature, depending on whether the circumference of the circle is less or greater than $2\pi r$. The term "curved space" should be abolished, because it is confusing and misleading. The moment we use the term "curved space", we make a declaration of adopting the extrinsic geometry. Another thing to note is that whether our real world is a curved space or Euclidean space is not the absolute truth of nature. It depends on the human convention of how we measure distance. In a nutshell, Riemannian geometry is just a geometry which is different from the Euclidean geometry, where "local government may stipulate different measuring rod standards". In the real world, whether the "space" is curved or flat is not the innate property of nature, but rather it is a human convention whether we choose to use Euclidean geometry or a different geometry in our life. Furthermore, the "space" itself literally means the nonexistence of matter, and "space" by itself is amorphous. The curved surfaces have material models. It does not make sense to call space curved or flat. Only the measurement of material bodies has a real meaning.

1.1 Extrinsic View of Curved Surfaces and Curved Spaces

Let $\mathbb{E}^3$ denote the 3-dimensional Euclidean space. A plane in $\mathbb{E}^3$ can be considered a subspace of $\mathbb{E}^3$, which is represented by a linear equation

$$ax + by + cz + d = 0,$$

where a, b, c, d are constants and $a^2 + b^2 + c^2 \neq 0$. A sphere (also called 2-sphere, denoted by S^2) of radius R can be described by equation

$$x^2 + y^2 + z^2 = R^2.$$

A cylinder of radius R can be described by equation

$$x^2 + y^2 = R^2.$$

A plane is considered a flat surface (2-dimensional Euclidean subspace $\mathbb{E}^2$ of $\mathbb{E}^3$), while cylinders and spheres are examples of curved surfaces in $\mathbb{E}^3$.

While studying the curves passing any point P on a surface, Euler finds that there are two directions (perpendicular to each other) in which their normal curvatures take maximal and minimal values κ_1 and κ_2. S. Germain (1831) defines

$$H = \frac{1}{2}(\kappa_1 + \kappa_2)$$

as the curvature of the surface at point P. This is known as the mean curvature. For a plane, $\kappa_1 = \kappa_2 = 0$, hence $H = 0$. For a sphere, $\kappa_1 = \kappa_2 = 1/R$, hence $H = 1/R$. For a cylinder, $\kappa_1 = 1/R$, $\kappa_2 = 0$, hence $H = 1/(2R)$. This mean curvature well captures our intuition of curvature as how much the surface deviates from a plane.

Now let us look at higher dimensions. Each point of the 4-dimensional Euclidean space $\mathbb{E}^4$ can be represented by coordinates (x, y, z, u). $\mathbb{E}^4$ has 2-dimensional metric subspaces (surfaces), and 3-dimensional metric subspaces (hypersurfaces). A linear equation

$$ax + by + cz + du + e = 0$$

represents a hyperplane. The equation

$$x^2 + y^2 + z^2 + u^2 = R^2$$

represents a hypersurface called hypersphere, or 3-sphere, denoted by S^3, which consists of all the points in $\mathbb{E}^4$ which have a constant distance R to the origin. The concept of curvature can be generalized to surfaces and hypersurfaces in $\mathbb{E}^4$. Intuition tells us that a hyperplane is flat but the hypersphere is curved. Therefore we expect that a hyperplane has zero curvature while a hypersphere has nonzero curvature. The curvature of a hypersurface captures how much it deviates from a hyperplane.

A curved hypersurface (like S^3) is a 3-dimensional subspace of the 4-dimensional space $\mathbb{E}^4$. We call it a "curved space". The curved space only makes sense as a subspace of a higher dimensional Euclidean space $\mathbb{E}^4$. The concept is a little harder than a 2-d curved surface, because it has higher dimensions, but still not too difficult. We live in a 3-d world $\mathbb{E}^3$. We have intuition of a curved surface like S^2 because we can see it and touch it. For a curved space or curved hypersurface, we do not have intuition about them, because we do not live in 4-d space $\mathbb{E}^4$. We can still study the properties of curved spaces by analytical method, with coordinates and equations.

1.2 Intrinsic View of Curved Surfaces due to Gauss

One question arises. Do we really live is a 3-d Euclidean space? We do not question the dimension, which is 3. We ask if it is possible that we live in a hypersphere S^3 instead of $\mathbb{E}^3$. What are the differences between S^3 and $\mathbb{E}^3$ as subspaces of $\mathbb{E}^4$? In a small neighborhood in S^3, the geometry is very similar to that of a small neighborhood in $\mathbb{E}^3$. Think in analogy: the surface of a tranquil lake looks flat to us, but we know it is part of the surface of the earth, which is a sphere. Is it possible to find out if we live in S^3 instead of $\mathbb{E}^3$, from inside our space?

The answer is yes. When it is difficult to infer the relationship of a hypersurface to $\mathbb{E}^4$, we always use an analogy in one dimension lower, which is the relationship of surfaces to space $\mathbb{E}^3$.

Think of a surface S. For simplicity, assume it is a sphere S^2. Suppose there is an earth mound in the shape of hemisphere on the ground and some ants live on the surface. Can the ants find out that they live on a curved surface by measurement on the surface (without leaving the surface)? The answer is yes. For this, we need to distinguish two types of distances. In Euclidean space $\mathbb{E}^3$, the distance between any two points $P_1 = (x_1, y_1, z_1)$ and $P_2 = (x_2, y_2, z_2)$ is defined to be

$$\rho \overset{\text{def}}{=} \sqrt{(x_2 - x_1)^2 + (y_2 - y_1)^2 + (z_2 - z_1)^2}. \tag{10.1}$$

We know this is the distance of the straight line segment P_1P_2, which is also the minimal length of any smooth curves in $\mathbb{E}^3$ connecting P_1 and P_2,

$$\rho = \inf_{\gamma \text{ in } \mathbb{E}^3} \int_{\substack{p_1 \\ \gamma}}^{p_2} ds.$$

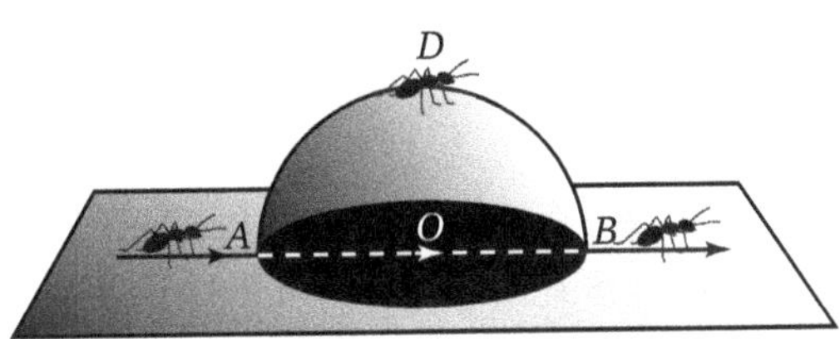

Figure 10.1   An ant on the ground with an earth mound

Suppose an ant is at point A and wants to go to the antipodal point B (Figure 10.1). The distance $\rho(A, B)$ between A and B is the diameter $\overline{AOB}$ of the sphere with a length of $2R$, which is the shortest travel distance possible for the ant. However, the ant has to dig a tunnel under the mound in order to reach B through this shortest path. What if such tunnels are not possible (say the mound has a concrete surface) and the ant has to stay on the surface all the time? The shortest possible path will be a half of a great circle $\overset{\frown}{ADB}$ with a length of πR. We call the shortest distance restricted *on the surface S* the *geodesic distance*, or the *intrinsic distance*,

$$\tilde{\rho} \overset{\text{def}}{=} \inf_{\gamma \text{ on } S} \int_{\substack{p_1 \\ \gamma}}^{p_2} ds. \tag{10.2}$$

We see $\rho \leq \tilde{\rho}$ because one may be able take a shortcut (tunnel) in 3-d space. If the ants are restricted to the surface, they cannot measure the distance ρ. The possible distance measurement accessible to them is only $\tilde{\rho}$, the intrinsic distance, or geodesic distance on the surface S. If we restrict ourselves to the concept of intrinsic distance $\tilde{\rho}$, and give up any reference to the distance ρ measured off the surface in space, we get intrinsic geometry of the surface. The properties studied in this method will be called intrinsic properties. If we bend the surface without stretching, we get another surface, but the intrinsic distance between any two points on the surface is preserved. These two surfaces will be deemed identical, and are called isometric to each other.

Let us define a curvature in the intrinsic way. The previously defined mean curvature H is not intrinsic because it refers to measurement outside the surface, which is out of reach to the inhabitants on the surface. At a point P on a surface S, draw a *geodesic circle*—points of equal intrinsic distance r to point P. r is called the radius. Let the circumference of the geodesic circle be $C(r)$. The intrinsic curvature, or Gaussian curvature is defined to be the limit[2]

$$K \overset{\text{def}}{=} \frac{3}{\pi} \lim_{r \to 0} \frac{2\pi r - C(r)}{r^3}.$$

Using this definition, the curvature of the plane is still zero because on a plane, $C(r) = 2\pi r$. The sphere has constant curvature $1/R^2$. So the Gaussian curvature also captures our intuition. What is the Gaussian curvature of the cylinder? It can be found that $K = 0$. This is different from the mean curvature H. It is counterintuitive at the first glance, because we have the intuition that the cylinder is a curved surface. But if we think more, it makes sense. We can roll a piece of paper (plane) into the shape of cylinder. This rolling is an isometric mapping because the intrinsic distance between any two points is preserved. As the Gaussian curvature is intrinsic (invariant under isometric mapping), the cylinder should have the same Gaussian curvature as the plane, which is zero. This means, the ants on a sphere are able to discover they live on a curved surface because on a sphere

---

[2]There are different definitions equivalent to each other.

$C(r) < 2\pi r$ and it has a positive curvature (Figure 10.2a), while the ants on a cylinder are not able to tell whether they live on a cylinder or a plane, because the intrinsic geometry of the two is the same (Figure 10.2b). The saddle surface has a negative curvature because on this surface $C(r) > 2\pi r$ (Figure 10.2c). Take a patch on the sphere. If we try to flatten it onto the plane, it has to be ripped, like the orange peel, because $C(r) < 2\pi r$ on the sphere (Figure 10.2d). If we take a patch from a cylinder and flatten it on the plane, it fits perfectly because $C(r) = 2\pi r$ on the cylinder (Figure 10.2e). If we take a patch from the saddle surface and flatten it on the plane, it will fold, because $C(r) > 2\pi r$ on the saddle surface (Figure 10.2f).

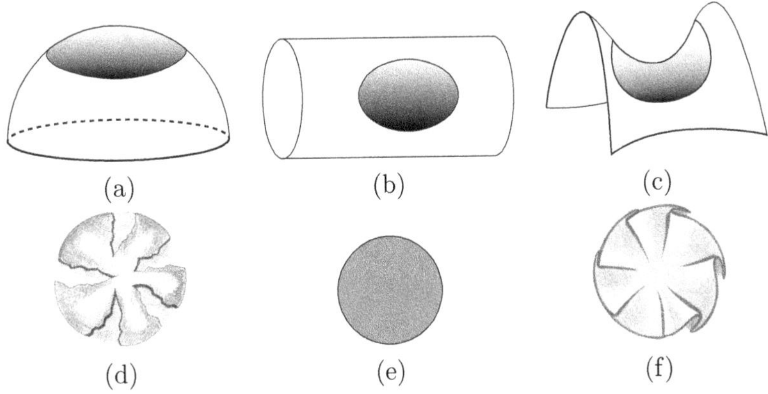

Figure 10.2   (a, d) A sphere (b, e) A cylinder (c, f) A saddle surface

If two points on the surface are infinitely close, the distance in Eq. 10.1 becomes

$$d\rho^2 = dx^2 + dy^2 + dz^2. \tag{10.3}$$

For two points P_1 and P_2 widely apart, the Euclidean distance $\rho(P_1, P_2)$ and the intrinsic distance $\tilde{\rho}(P_1, P_2)$ on the surface are not equal in general, with $\rho(P_1, P_2) \leq \tilde{\rho}(P_1, P_2)$. When P_1 and P_2 are infinitely close to each other, the Euclidean distance $d\rho$ and the intrinsic distance $d\tilde{\rho}$ become equal. Without a need to distinguish them, we will just use ds, with $ds \stackrel{\text{def}}{=} d\tilde{\rho} = d\rho$, which is called the line element, and we have

$$ds^2 \stackrel{\text{def}}{=} d\tilde{\rho}^2 = d\rho^2 = dx^2 + dy^2 + dz^2. \tag{10.4}$$

We say the intrinsic distance $d\tilde{\rho}$ is inherited from, or induced by the Euclidean distance $d\rho$. The three coordinates x, y, z for points on the surface are not independent. They satisfy a constraint, which is the equation of the surface

$$f(x, y, z) = 0. \tag{10.5}$$

We may use two independent parameters u and v to represent the surface in the form of parametric equations

$$x = x(u, v),$$
$$y = y(u, v), \tag{10.6}$$
$$z = z(u, v).$$

For the example of a sphere, we can use the polar coordinates θ and φ, with θ being the latitude and φ the longitude,

$$x = R \cos\theta \cos\varphi,$$
$$y = R \cos\theta \sin\varphi,$$
$$z = R \sin\theta.$$

Substitute this into Eq. 10.4, we obtain

$$ds^2 = R^2 \left(d\theta^2 + \cos^2\theta d\varphi^2 \right).$$

In general, the square of the line element of any surface is a positive-definite differential quadratic form

$$ds^2 = E(u, v)du^2 + 2F(u, v)dudv + G(u, v)dv^2.$$

If we use indexed parameters x_1 for u and x_2 for v, then

$$ds^2 = \sum_{i=1}^{2} g_{ij}(x_1, x_2)dx_i dx_j, \tag{10.7}$$

where $g_{ij}(x_1, x_2)$ is a symmetric tensor field called the metric tensor (field). The old name is the first fundamental form, which is a quadratic form. Gauss showed that all the intrinsic properties of the surface can be inferred from this metric tensor (field).

1.3   Riemann's Generalization of the Intrinsic Geometry

Riemann generalized the intrinsic geometry of Gauss even further in two aspects.

(1) There is no need for the definition of the surface with Eq. 10.5 or Eq. 10.6.[3] We can take the metric tensor $ds^2 = \sum_{i=1}^{2} g_{ij} dx_i dx_j$ in Eq. 10.7 as the starting point and the definition of the surface.

(2) Generalize the surface from dimension 2 to arbitrary dimension n, meaning to use a symmetric positive-definite differential bilinear form $ds^2 = \sum_{i=1}^{n} g_{ij} dx_i dx_j$ as the starting point and definition of the generalized space, which is now called a Riemannian manifold.

Putting them together, the following are the two fundamental ideas of Riemannian geometry, or Riemannian manifold:

(R1) The dimension of the space is any integer $n > 0$ (generalization to higher dimensions).

(R2) The (length measurement of the) space is defined by the inner product in the tangent space (metric tensor)

$$ds^2 = \sum_{i=1}^{n} g_{ij} dx_i dx_j.$$

Informally ds is interpreted as the infinitesimal distance between two nearby points.

The key idea of the Riemannian manifold is the metric tensor g_{ij}, which may vary from point to point in space. Effectively, this is to allow different length measuring standards in different locations in space. If the metric tensor g_{ij} is a constant everywhere (not necessarily orthogonal or diagonal), it is a Euclidean space $\mathbb{E}^n$. When g_{ij} varies from place to place, it is a Riemannian manifold. There are cases that g_{ij} varies from place to place, but it is still Euclidean space (at least locally). This is the case of Euclidean space $\mathbb{E}^n$ in the disguise of curvilinear coordinates. In such a case, with a coordinate transformation, the metric tensor g_{ij} can be made constant (and diagonal).

It is a misconception that Riemannian geometry is the geometry of curved spaces of higher dimensions ≥ 3. Of course 2-dimensional curved surfaces embedded in $\mathbb{E}^3$ are simple examples of 2-dimensional Riemannian manifolds, but we may have 2-dimensional Riemannian manifolds which are not surfaces in $\mathbb{E}^3$. These are called abstract surfaces. Examples include the hyperbolic plane and the elliptic plane. The hyperbolic plane is defined by the metric

$$ds^2 = \frac{1}{x_2^2}(dx_1^2 + dx_2^2), \ x_2 > 0.$$

---

[3]This is considered the embedding of the surface into Euclidean space $\mathbb{E}^3$, and the complete intrinsic geometry is to ignore the embedding.

It is an abstract surface, but not a surface embedded in $\mathbb{E}^3$. This is the Poincaré half-plane model, but there are many other different parameterizations which describe the same abstract surface. M. do Carmo [(1976)] has a nice discussion of abstract surfaces.

We have seen that the concept of "curved space" is not hard to understand at all if we take the extrinsic view, meaning to view it as a hypersurface (like $x^2 + y^2 + z^2 + u^2 = R^2$) embedded in a higher dimensional Euclidean space $\mathbb{E}^n$. What is hard is the intrinsic view. That is, we can study the curved surface as an entity of its own right, from inside, without the need of embedding into a higher dimensional ambient Euclidean space.

Remark. Conventionalism: What is "straight", "flat", and what is "curved", are relative to human conventions. If we adopt the Lobachevsky geometry, the Euclidean plane is curved. Suppose we live in Lobachevsky space. The Lobachevsky plane is just flat if we adopt the Lobachevsky geometry as the government stipulated official geometry. A curved surface called a horosphere in Lobachevsky space is basically a Euclidean plane. So the Euclidean plane is a curved surface in Lobachevsky geometry. This means that the concepts of "curved", "straight", "flat" are all relative to the official geometry adopted by convention, which dictates the measurement standard.

§2. What Is a Tangent Space Exactly?

2.1   Extrinsic View Is Easy

Tangent spaces of a differentiable manifold are the high dimensional generalization of tangent planes of a surface in $\mathbb{R}^3$. However, in intrinsic geometry, things become more difficult, because the ambient Euclidean space is gone.

Let us first have a look at the 2-sphere S^2 in $\mathbb{R}^3$ defined as a subset of points $(x_1, x_2, x_3) \in \mathbb{R}^3$: $x_1^2 + x_2^2 + x_3^2 = 1$ (Figure 10.3).

At any point $A = (a_1, a_2, a_3)$ on the sphere, $\mathbf{a} = \overrightarrow{OA} = (a_1, a_2, a_3)$ is a normal vector of the surface. Let $\xi = (\xi_1, \xi_2, \xi_3) \in \mathbb{R}^3$ be a point P in the tangent plane at A. The vector

$$\overrightarrow{AP} = \overrightarrow{OP} - \overrightarrow{OA} = (\xi_1 - a_1, \xi_2 - a_2, \xi_3 - a_3)$$

is called a tangent vector. Any tangent vector in the tangent plane is perpendicular to the normal vector $\mathbf{a}$ and hence $\mathbf{a} \cdot \overrightarrow{AP} = 0$. This is the equation of the tangent plane,

$$a_1(\xi_1 - a_1) + a_2(\xi_2 - a_2) + a_3(\xi_3 - a_3) = 0.$$

This can be simplified to

$$a_1\xi_1 + a_2\xi_2 + a_3\xi_3 = 1.$$

For the tangent plane, if we move the origin to A, and a point in the tangent plane has coordinates $\xi' = (\xi_1', \xi_2', \xi_3')$, then the equation of the tangent plane is

$$a_1\xi_1' + a_2\xi_2' + a_3\xi_3' = 0.$$

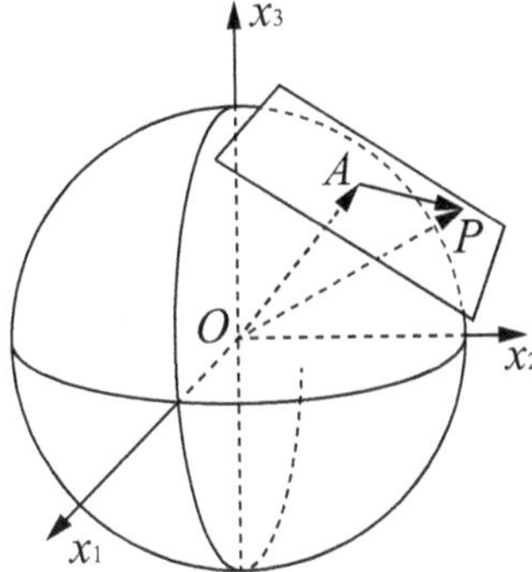

Figure 10.3   Tangent plane of a sphere.

When a surface in $\mathbb{R}^3$ is generalized to a hypersurface (differentiable manifold) in $\mathbb{R}^n$, the concept of tangent plane is generalized to tangent space, or tangent hyperplane. For example, we can have a 3-sphere S^3 defined as a subset of points $(x_1, x_2, x_3, x_4) \in \mathbb{R}^4$,

$$x_1^2 + x_2^2 + x_3^2 + x_4^2 = 1.$$

At any point $A = (a_1, a_2, a_3, a_4)$ on S^3, $\mathbf{a} = \overrightarrow{OA} = (a_1, a_2, a_3, a_4)$ is a normal vector. We have a 3-dimensional hyperplane in $\mathbb{R}^4$, called the tangent space of S^3 at point A. Let $\xi = (\xi_1, \xi_2, \xi_3, \xi_4)$ be a point P in the tangent space at A. The vector

$$\overrightarrow{AP} = \overrightarrow{OP} - \overrightarrow{OA} = (\xi_1 - a_1, \xi_2 - a_2, \xi_3 - a_3, \xi_4 - a_4)$$

is called a tangent vector. Any tangent vector in the tangent space is perpendicular to the normal vector $\mathbf{a}$ and hence $\mathbf{a} \cdot \overrightarrow{AP} = 0$. This is the equation of the tangent space, which can be simplified to

$$a_1\xi_1 + a_2\xi_2 + a_3\xi_3 + a_4\xi_4 = 1.$$

It is a 3-dimensional hyperplane in $\mathbb{R}^4$. If we move the origin of space to A, the equation of the tangent space becomes

$$a_1 \xi_1' + a_2 \xi_2' + a_3 \xi_3' + a_4 \xi_4' = 0$$

in the new coordinates.

2.2   Intrinsic View Is More Difficult

The above discussion is the traditional extrinsic view before Riemann. Riemann steered the direction of differential geometry toward higher dimensions in a completely intrinsic way. In the intrinsic approach, we have abandoned the Euclidean space as a container for the differentiable manifold. The normal direction or normal vector of the differentiable manifold at a point is forever gone.

Can we still define the tangent space of a differentiable manifold? It turns out that we can still manage to recover the concept of tangent space without going out of the manifold, but we just need to look at them from another perspective.

We start with surfaces in $\mathbb{R}^3$. Higher dimensions are just similar. This alternative perspective is something that we have already happily accepted. That is, in an "infinitesimal neighborhood", the surface is approximately flat, and the tangent plane is the linear approximation of the surface. This agrees with our everyday experience on earth. When you sit on a boat on a tranquil lake, the surface of the lake looks flat, although you know that the lake is part of the surface of a sphere. The tangent plane is in fact an infinite plane, but we can linearly extend the "infinitesimal neighborhood" of the surface to make it an infinite tangent plane.

Observe the curves passing a point p on the surface in $\mathbb{R}^3$ (Figure 10.4). Hereafter when we say a curve, we always mean a smooth curve. Each curve has a velocity vector which is tangent to the curve at p. The tangent plane at p consists of all these velocity vectors tangent to various curves passing p. We use a position vector $\mathbf{r} = (x, y, z) \in \mathbb{R}^3$ to represent a point on the surface. A parameterized curve is a smooth mapping $\gamma : \mathbb{R} \to M$, where M is the surface. In fact we are only concerned with an infinitesimal segment of the curve centered at p. Let $p = \gamma(0) = \mathbf{r}(0)$, and $\mathbf{r}(t) = \gamma(t) = (x(t), y(t), z(t))$ be a nearby point on the curve.

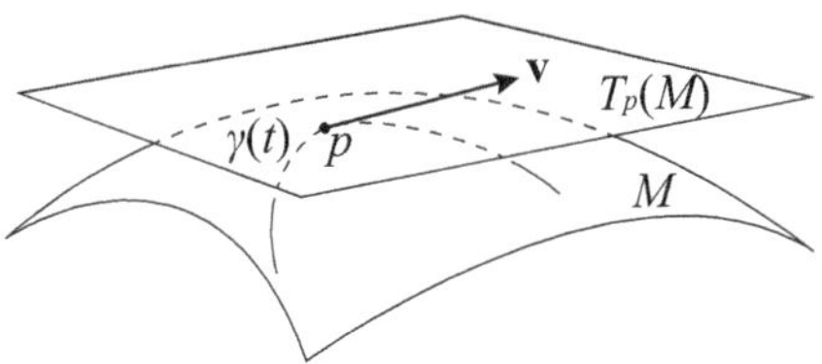

Figure 10.4   Tangent vectors and tangent plane.

The tangent vector at point p which is tangent to curve γ is the velocity vector

$$\mathbf{v} = \frac{d\mathbf{r}(t)}{dt}\Big|_{t=0} = \lim_{t \to 0} \frac{\mathbf{r}(t) - \mathbf{r}(0)}{t}.$$

The tangent vector $\mathbf{v}$ thus defined is a vector in $\mathbb{R}^3$. All the tangent vectors at point p which is tangent to some curve passing p form a plane, which is the tangent plane at p.

For an abstract manifold, we no longer have the surrounding space $\mathbb{R}^3$ or $\mathbb{R}^{n+1}$. We can no longer define a tangent vector living in $\mathbb{R}^3$, but we can use the curves $\gamma : \mathbb{R} \to M$ on the manifold. Let $p' = \gamma(t)$ be a nearby point of point $p = \gamma(0)$. We cannot use $\lim_{t \to 0} \frac{\gamma(t) - \gamma(0)}{t}$ or $\frac{d\gamma(t)}{dt}$ anymore. On the abstract manifold, subtraction of two points p and p', as well as the scalar multiplication by a real number $\frac{1}{t}$ is undefined!

Observe a family of curves that are all tangent to each other at point p. What they have in common is that their velocities at p are all in the same direction. How do we define that the two curves have the same velocity at p? We realized that although we have lost the surrounding Euclidean space $\mathbb{R}^{n+1}$, we still have another Euclidean space to our assistance. That is, a manifold is locally modeled on a Euclidean space $\mathbb{R}^n$, which provides the local coordinate system.

Intrinsically, without reference to the ambient Euclidean space, a tangent vector can be defined as an equivalent class of curves. The tangent plane is then the set of all tangent vectors at the same point. What is the intuition of defining a tangent plane intrinsically? Intuitively, a very small piece (infinitesimal neighborhood) of the surface itself is the tangent plane. For example, the tranquil surface of a lake can be taken as the tangent plane of the earth at the center point of the lake. Of course this tangent plane can be extended infinitely in our imagination, and its properties deviates from that of the surface when it is too far off the surface, but the part of the tangent plane which has physical interaction with the surface (manifold) is that infinitesimal region.

For a surface embedded in $\mathbb{E}^3$, there is also a normal vector at each point. This normal vector is embedded in $\mathbb{E}^3$ and is an extrinsic concept. In the intrinsic geometry of Gauss, the normal vector is still in natural existence. It is just that his final result does not involve the normal vectors. The surfaces are still defined as a point set with each point having three coordinates (x, y, z). In a Riemannian abstract surface (as well as in higher dimensions), the concept of normal vector is forever gone.

In higher dimensions, the concept of tangent plane becomes tangent space. A tangent space is defined for any differentiable manifold. If the dimension of the differentiable manifold is n, then the tangent space is an n-dimensional vector space. Note that the concept of length or distance does not exist in either that tangent space, or the differentiable manifold. If the tangent space is endowed with additional metric structure (inner product), it becomes a Euclidean space, and the differentiable manifold becomes a Riemannian manifold (see more in Sec. 4).

§3. Tensor Transformation Laws Revisited

In an infinitesimal neighborhood of a point p of the differentiable manifold M, we can use local coordinates $(x^1, \ldots, x^n)$. The tangent vectors which are tangent to the coordinate lines passing p form a basis of the tangent space $T_p(M)$, which is called the natural basis. Given a scalar field $f : M \to \mathbb{R}$, and a tangent vector $\mathbf{v} = (v^1, \ldots, v^n)$ under the natural basis, the directional directive $\nabla_{\mathbf{v}} f|_p$ of f in the direction of $\mathbf{v}$ can be defined. A tangent vector is no longer a protruding arrow, and it only serves the purpose of directional derivatives. In fact, in some books a tangent vector is *identified*, or *defined*, as a directional derivative operator $\nabla_{\mathbf{v}}$ through a set of axioms. Under the natural basis,

$$\nabla_{\mathbf{v}} = \sum_{i=1}^{n} v^i \frac{\partial}{\partial x^i}.$$

The set of operators $(\frac{\partial}{\partial x^1}, \ldots, \frac{\partial}{\partial x^n})$ can be identified as the natural basis. If we have a local coordinate change,

$$\bar{x}^i = \bar{x}^i(x^1, \ldots, x^n), \ i = 1, \ldots, n, \tag{10.8}$$

the new natural basis becomes $(\frac{\partial}{\partial \bar{x}^1}, \ldots, \frac{\partial}{\partial \bar{x}^n})$. Therefore,

$$\nabla_{\mathbf{v}} f = \sum_{k=1}^{n} v^k \frac{\partial f}{\partial x^k}$$

$$= \sum_{k=1}^{n} \sum_{i=1}^{n} v^k \frac{\partial \bar{x}^i}{\partial x^k} \frac{\partial f}{\partial \bar{x}^i}.$$

This can also be written as

$$\nabla_{\mathbf{v}} = \sum_{k=1}^{n} v^k \frac{\partial}{\partial x^k}$$

$$= \sum_{k=1}^{n} \sum_{i=1}^{n} v^k \frac{\partial \bar{x}^i}{\partial x^k} \frac{\partial}{\partial \bar{x}^i}$$

$$= \sum_{i=1}^{n} \left(\sum_{k=1}^{n} \frac{\partial \bar{x}^i}{\partial x^k} v^k \right) \frac{\partial}{\partial \bar{x}^i}.$$

The components of the tangent vector $\mathbf{v}$ in the new basis are

$$\bar{v}^i = \sum_{k=1}^{n} \frac{\partial \bar{x}^i}{\partial x^k} v^k. \tag{10.9}$$

This is the coordinate transformation law of a contravariant vector.

We can make basis change in the tangent space more explicit. Let us denote the old basis by

$$\mathbf{e}_i = \frac{\partial}{\partial x^i},$$

and denote the new basis by

$$\bar{\mathbf{e}}_i = \frac{\partial}{\partial \bar{x}^i}.$$

We know

$$\bar{\mathbf{e}}_i = \frac{\partial}{\partial \bar{x}^i}$$

$$= \sum_{k=1}^{n} \frac{\partial x^k}{\partial \bar{x}} \frac{\partial}{\partial x^k} \tag{10.10}$$

$$= \sum_{k=1}^{n} \frac{\partial x^k}{\partial \bar{x}^i} \mathbf{e}_k.$$

This means that the local coordinate change in Eq. 10.8 induces a basis change in the tangent space, and the coordinate change of a vector in

Eq. 10.9 is the result of this basis change. The tangent vectors are called contravariant vectors because the transformation is represented by the matrix $\left[\dfrac{\partial \bar{x}^i}{\partial x^k}\right]$, which is the transpose of the inverse matrix ("backward transformation") for the basis change matrix $\left[\dfrac{\partial x^k}{\partial \bar{x}^i}\right]$ ("forward transformation").

We can construct tensor spaces $T_p(M) \otimes T_p(M)$, $T_p^*(M) \otimes T_p^*(M)$ and $T_p(M) \otimes T_p^*(M)$. The coordinates of a contravariant tensor $A \in T_p(M) \otimes T_p(M)$ will transform according to the following,

$$\bar{\xi}^{ij} = \sum_{r,s=1}^{n} \frac{\partial \bar{x}^i}{\partial x^r} \frac{\partial \bar{x}^j}{\partial x^s} \xi^{rs},$$

which we recognize as the transformation law as part of the old-fashioned definition of tensor (field) due to Ricci, as in Definition 1 in Chap. 1.

§4. What Are the Differences?
Differentiable Manifold vs. Riemannian Manifold

We often see an intuitive description: a differentiable manifold is the generalization of smooth surfaces to higher dimensions; locally a differentiable manifold "looks like" a Euclidean space. However, because this explanation is an analogy, it is not precise and may lead to misconceptions. This generalization is precisely in what way? What is the exact meaning of "a differentiable manifold locally *looks like* a Euclidean space"?

A differentiable manifold is indeed a generalization of smooth surfaces, but along the way of generalization, we have discarded the concept of distance, which we do have in the differential geometry of surfaces. So, more precisely, locally a differentiable manifold looks like a vector space, instead of a Euclidean space. Exterior derivative and Lie derivative can be defined on a differentiable manifold, but without additional structures (like distance derived from the metric tensor), the concept of curvature does not apply to differentiable manifolds, nor does covariant derivative.

The structure point of view is much needed to understand modern mathematical concepts (see Appendix 2). The Riemannian manifold has an additional structure, the metric tensor, through which we can define distance and angle. A Riemannian manifold can also be called the metric manifold. The Riemannian manifold is the generalization of surfaces and it keeps the concepts of distance and angle. Locally it looks like a Euclidean space.

A differentiable manifold, (and topological manifold as well), is not considered the "curved space". Two topological manifolds are considered the same if one can be "morphed" to the other via a topological transformation (homeomorphism)—continuous stretching. Two differentiable manifolds are considered the same if one can be "morphed" to the other via a diffeomorphism—smooth stretching. Two Riemannian manifolds are considered the same if one can be "morphed" to the other via an isometric transformation—bending without stretching which preserves the intrinsic distance between any two points. The study of differentiable manifolds without additional structures belongs to differential topology, while the study of Riemannian manifolds is considered differential geometry.

We cannot talk about whether a differentiable manifold (without additional structures) is curved or not, or the curvature of a differentiable manifold, because it cannot be defined. The same differentiable manifold can be equipped with different metric structures so that they may have different curvatures. Take the torus as an example of a differentiable manifold. We cannot say it is curved or not. Figure 10.5(a) shows an ordinary torus T^2 (embedded in $\mathbb{E}^3$).

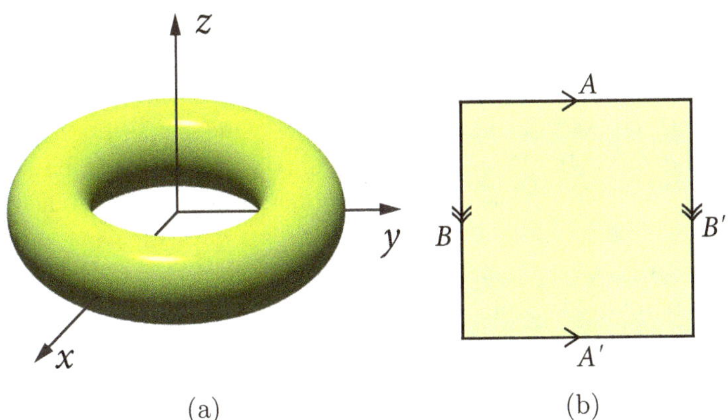

Figure 10.5   The ordinary torus by gluing and stretching

The torus can be described by the equation

$$\left(\sqrt{x^2 + y^2} - a\right)^2 + z^2 = b^2,$$

or parametric equations

$$x = (a + b\cos\theta)\cos\varphi,$$
$$y = (a + b\cos\theta)\sin\varphi,$$
$$z = b\sin\theta.$$

A plane passing through the z-axis intersects the torus with two small circles of radius b in the vertical plane. The locus of the center of such vertical circles is a circle of radius a on the x-y plane. The Gaussian curvature depends on θ only and is independent of φ,

$$K = \frac{\cos\theta}{b\,(a + b\cos\theta)}.$$

On the very top and at the very bottom of the torus ($z = \pm b$, $\theta = \pm\pi/2$) are the circles C_1 and C_2 with radius a and the equation

$$x^2 + y^2 = a^2.$$

These two circles divide the torus into the outer part with $\sqrt{x^2 + y^2} > a$ and the inner part with $\sqrt{x^2 + y^2} < a$. The Gaussian curvature on C_1 and C_2 is zero, while positive on the outer part and negative on the inner part.

The topology of the torus is $S^1 \times S^1$, with S^1 being a circle. We can make a torus by gluing the edges of a square (or rectangle), shown in Figure 10.5(b). First we glue the edges A and A' to form a cylinder. The paper is bent with no stretch. Then we bend the cylinder and glue the edges B and B' together. In this second bending, the surface must be stretched. Paper will not work. It needs to be a rubber sheet to be stretched.

It is possible to make a flat torus. We start with the same rectangular sheet, but we glue the opposite edges without bending or stretching. Such a torus cannot live in $\mathbb{E}^3$ but can live in $\mathbb{E}^4$. In $\mathbb{E}^4$ with coordinates (x, y, z, u), its equations are

$$\begin{aligned} x^2 + y^2 &= a^2, \\ z^2 + u^2 &= b^2, \end{aligned} \tag{10.11}$$

or in parametric form

$$\begin{aligned} x &= a\cos\varphi, \\ y &= a\sin\varphi, \\ z &= b\cos\theta, \\ u &= b\sin\theta. \end{aligned} \tag{10.12}$$

The geometry is the Euclidean geometry with line element

$$ds^2 = dx^2 + dy^2 + dz^2 + du^2$$
$$= a^2 d\varphi^2 + b^2 d\theta^2.$$

If we define $X = a\varphi$ and $Y = b\theta$, then

$$ds^2 = dX^2 + dY^2,$$

which is the same as the first fundamental form of the Euclidean plane.

Flat torus is another example of an abstract surface, or 2-d Riemannian manifold, which is not a surface in $\mathbb{E}^3$ (cannot be isometrically embedded in $\mathbb{E}^3$). Eq. 10.12 is a straightforward embedding of the flat torus in $\mathbb{E}^4$. Recall that the hyperbolic plane is another abstract surface which cannot be isometrically embedded in $\mathbb{E}^3$ (Sec. 1). D. Blanuša [(1955)] constructed an isometric embedding of the hyperbolic plane in $\mathbb{E}^6$.

We can visualize a realization of the flat torus in real world, in fact in $\mathbb{E}^2$. That will be the screen of some video games, for example, a 2-d game of shooting airplanes (Figure 10.6). When the airplane hits the edge A, it does not go out, or vanish, but emerges at edge A'.

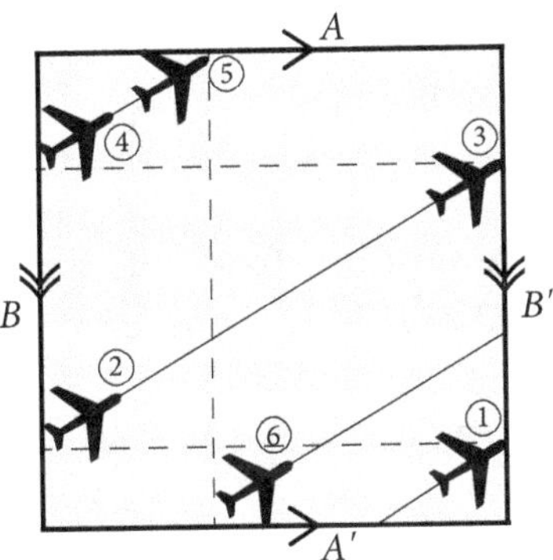

Figure 10.6   A flat torus by gluing ("electronically") without stretching

The ordinary torus and the flat torus have the same topology. One can be morphed onto another via a homeomorphism. They even have the same differential structure, meaning this can even be a smooth morphing (diffeomorphism). However, they have different metric structures (curvatures). The ordinary torus has nonzero Gaussian curvatures, while the curvature of the flat torus is zero everywhere. This means the morphing between the two cannot be isometric. In other words, one cannot be morphed onto another without stretching.

For another example, let us look at the distinction between a Euclidean space and a vector space. A vector space is defined in Definition 11 in Chap. 1. It has two operations, the addition of two vectors and the scalar-vector multiplication. A vector space endowed with an inner product is called a Euclidean space. We denote the n-dimensional real vector space by $\mathbb{R}^n$ and the Euclidean space by $\mathbb{E}^n$. The difference between the two is subtle. The Euclidean space has an additional structure, the inner product, by which we can define length and angle. In a vector space, the length, or magnitude of a vector is not defined. This means in a vector space, by convention, we are not allowed to talk about the length of a vector.

Sherman Stein published a book entitled *Mathematics: the Man-Made Universe*. The book title itself reflects the author's view, as he further states in the preface: "Mathematics, on the other hand, is completely the work of man." We view all the mathematical entities as human constructions. We do not take it for granted that the concept of "distance" is natural existence. If we build it there, as in Euclidean space, then it has the concept of distance. If we decide to omit it, then it is absent, as in a vector space, and we refrain from talking about distance. It is meaningless to debate whether the concept of distance is in natural existence. It is just a convention, or an agreement between the author and the readers whether we decide to talk about it or not. This way, we can focus on the important concepts.

§5. How Can Riemannian Geometry Be Applied to the Real World? —Conventionalism

While mathematicians have their liberty to invent their abstract mathematical theories and build a man-made universe, physicists have their liberty to pick and apply whatever mathematical theory to the real world, as long as they can find a physical model for the mathematical structures.

Can Riemannian geometry be applied to the real world? One may think of general relativity, but what is applied in general relativity is the pseudo-Riemannian geometry, where part of spacetime is time. Pseudo-Riemannian geometry is not really geometry, where you cannot talk about lengths and angles (see more in Sec. 6).

Can Riemannian geometry be applied to our 3-dimensional space with the concepts of distance and angle? Could the space we live in be really curved?

If we live in a 3-dimensional Riemannian manifold with constant curvature, things seem to be fine. A hyperbolic space is a Riemannian manifold with constant negative curvature, while an elliptic space is a Riemannian manifold with constant positive curvature. The third possibility is the Euclidean space with constant zero curvature.) In a public lecture delivered at the University of Singapore on June 27, 1980, Shiing-Shen Chern said, "In spite of the success of Euclid it is not clear why our space should be Euclidean." My opinion is, our space being Euclidean is not the innate property of nature, but it is the result of the human convention of length measurement standard we have chosen.

One concern of applying Riemannian geometry is that, if the curvature of the Riemannian manifold is not a constant everywhere, then there will not be the concept of rigid bodies—namely, such rigid bodies cannot exist, because a body moving from one place to another place with different curvature will have to be distorted, not by physical forces, but by "space"! (This literally does not make sense.)

My opinion is that Riemannian geometry (even with varying curvature) can be applied to describe the real world. This is because, whether the space is curved or flat is not the innate property of nature, but rather the human convention of length measurement standard. We shall show this with an example of the geometry of the surface of the earth (a sphere),

$$x = R\cos\theta\cos\varphi,$$

$$y = R\cos\theta\sin\varphi,$$

$$z = R\sin\theta, \qquad \left(-\frac{\pi}{2} < \theta < \frac{\pi}{2}, 0 < \varphi < 2\pi\right)$$

where R is the radius of the earth, θ is the latitude and φ is the longitude. If we stipulate the metric tensor as

$$ds^2 = R^2 \left(d\theta^2 + \cos^2\theta d\varphi^2\right), \tag{10.13}$$

we obtain the ordinary spherical geometry with a constant curvature of $1/R^2$. How do we implement this metric? We keep the *international prototype meter* (IPM), a standard meter in the form of a platinum alloy bar in Paris. When we need to measure length in a location, say Oslo, Norway, we move the IPM bar to Oslo and use it to measure (practically the IPM bar is duplicated and the duplicates are moved to different countries). This is the length measurement standard or convention that implements the metric in Eq. 10.13.

However, we could as well adopt a different convention of Riemannian metric

$$ds^2 = R^2 \left(\sec^2 \theta d\theta^2 + d\varphi^2\right).\tag{10.14}$$

If we make changes of variables,

$$x = R\varphi,\tag{10.15}$$

$$y = R\ln\left(\tan\theta + \sec\theta\right),\tag{10.16}$$

and substitute them into Eq. 10.14, we obtain

$$ds^2 = dx^2 + dy^2.$$

It is easy to see that the Gaussian curvature of this metric is zero. This means the earth is flat and we can use Euclidean plane geometry on the surface of the earth! This should not be a surprising fact. This shows whether the earth is curved or flat is not an absolute truth. It is the result of human convention of length measurement standard. In fact Eq. 10.14 is just the metric for the Mercator map (Figure 10.7), while the map is scaled down by proportion. We may call Eq. 10.14 the Mercator metric.

Figure 10.7   The Mercator metric and the flat earth

How do we implement the metric in Eq. 10.14? Think of the following fictitious scenario: the International Bureau of Weights and Measures

(BIPM) has convened and agreed to switch to the Riemannian geometry with Mercator metric starting January 1, 2022. What changes will this switch bring about? Each city government will keep a local standard meter in the form of a platinum bar! Each local standard meter bar should comply with the metric at latitude θ and longitude φ. This will not affect the geometry locally within each city, except each city will have a new length unit, and the government should promulgate the conversion rate between the new meter and the old meter. The local geometry within each city is still Euclidean. However, the geometry in the large will be changed and it will result in the following unusual phenomena:

(1) The area of Greenland is larger than the continent of South America. (You may have learned the fact that "Greenland is about 1/8 the area of South America". The fact is, both statements "Greenland is larger than South America" and "Greenland is 1/8 of South America" are facts. There are no absolute facts. There are conventions in facts and there are conventions in physical laws.)

(2) The Pythagorean theorem holds even for any big intercontinental triangles.

(3) All parallels of latitude have equal length.

(4) The geodesics on the surface of the earth are straight lines, which measure the shortest distance (remember, each portion of the line passing any city is measured using the local standard meter bar).

(5) However, airplanes will choose to fly on curved paths rather than geodesics (straight lines)[4], because they use less fuel even if they fly longer distances than geodesic distances. You might protest, "But the Mercator distance is not the true distance!" Well, there is no true distance. Your "true distance" is nothing but a distance by a different convention which you are accustomed to and is possibly more convenient.

(6) People in Oslo, Norway (near latitude $\theta = 60°$, scale factor $k = \sec\theta = 2$) are about two times as tall[5] as people in Singapore (near the equator $\theta = 0$, scale factor $k = \sec\theta = 1$). If a person travels from Singapore to Oslo, he grows two times as tall, but when a person travels from Oslo to Singapore, his height is reduced to half (length contraction)! Houses and cars in Oslo are all two times as large as those in Singapore, but local people do not feel happier about this because their body sizes are two

---

[4]With a few exceptions like the equator and the meridians.

[5]A person's height is measured when lying down, to stay in the idealized 2-dimensional geometry on the sphere.

times as large too. This length dilation and contraction even happen to platinum bars. Hence there do not exist any rigid materials. Suppose an engineer flies from Singapore to Oslo, caring with him a wrist watch, some Singapore money and a measuring tape. Upon arrival, he needs to adjust his watch to local time and exchange his money to local currency. This is what all of us do with international travel, but with the adoption of the new Riemannian geometry, he also needs to buy a new measuring tape in a local store, because the measuring tape he carried with him from Singapore has grown twice as long and can no longer be used for length measurement in Oslo!

Note that (1) through (4) are facts in geometry. (5) and (6) are the laws of physics and biology, which are not governed by geometry, but what geometry we adopt certainly plays a role in the laws of physics. With the ordinary 3-d Euclidean geometry, the platinum bar is rigid. In this relationship between geometric laws (Euclidean geometry) and physical laws (that the platinum bar is rigid), which is the cause and which is the effect? Actually it is the convention (of the physical law that the platinum bar is rigid) that determines the 3-dimensional geometry around the earth is flat Euclidean (and the surface of the earth is curved).

Think about the traditional approach, why is our geometry Euclidean, and why is the Paris IPM bar rigid? One fact answers both questions: we use the same IPM bar to measure lengths in all locations. When the Paris IPM bar is moved to Singapore or Oslo, we use the same bar as the unit of length in all these locations. May the length of the bar change due to the humidity in Singapore or the cold temperature in Oslo? No. Why? "Because I said so!" This means it is a stipulation, or convention. By definition, or convention, the Paris IPM bar does not change length. On the contrary, if we adopt the Riemannian geometry with Mercator metric, the Paris IPM bar changes length when moved to Singapore or Oslo, because we have different standard meters at different locations.

We may extend the Mercator metric in Eq. 10.14 to 3-dimensional space,

$$ds^2 = dr^2 + r^2 \sec^2 \theta d\theta^2 + r^2 d\varphi^2.$$

This is a curved space! We do not need to go to remote galaxies to see the reality of curved space. It is right here around us on earth! This corroborates our remarks in Sec. 1: curved space means nothing but a space where we adopt geometric laws which are different from Euclidean geometry. Life will be normal and everything will be normal, if we adopt this Riemannian geometry with Mercator metric. This is why Poincaré says: "One geometry is not more true than another; it can only be more convenient."

In Sec. 1, we used an analogy of ants on a curved surface to explain the idea of curved space, or a possible application of Riemannian geometry. The ants confined on the surface are not able to have a view from above the surface, but a human, as a superbeing, can look over their universe and see from outside that the space of their world is a curved surface. The ants can only discover this fact from inside, by measuring lengths using a taut string. However, there is a difference between the 2-dimensional world for ants in the analogy and the possible 3-dimensional curved space that we humans might live in. In that example (see Figure 10.1), the surface as the 2-dimensional space for ants is a material surface. The ants are constrained on that surface (which is their free space) by physical forces (like gravity) in the 3-d world. In their 2-dimensional curved space, there is no distinction what is mater and what is free space (void), but we do have this distinction in our 3-dimensional world. To make the analogy of the 2-dimensional world for the ants more similar to our 3-dimensional world, we modify the model as follows (Figure 10.8). The world for ants are still 2-dimensional (with zero thickness). However, we distinguish what is 2-dimensional matter (2-d Earth, 2-d Moon, 2-d tower, 2-d ants) and what is space (void, or absence of matter). The Earth and the Moon are 2-dimensional material entities on the surface of a sphere S^2. The ants are also material entities. There can also be towers erected on Earth. What is different from Figure 10.1 is that we make clear that other than these material entities, the rest of the world is filled with "space", which means void. For the 2-d Earth, a human as a superbeing can see clearly that it is part of a sphere S^2, because it is material existence and it is real. When an ant rides a rocket to go from the Earth to the Moon, it enters the space—the void. It is not enforced by physical laws, or supported by material constraints to stay on the surface of S^2. That is to say, whether the space the ants live in—the real space, meaning the void—is curved or flat does not have a meaning. The ant does not know, and even a human watching from above in the privileged 3-d world cannot see, whether the ant is staying on S^2. The void part of S^2 (as opposed to the material part of S^2) is only imaginary and invisible even to the human as a superbeing. We are not even sure that the ant is staying on any surface during this travel in the void space. During the ant's journey in the void to the 2-d Moon, we are not even sure, and it has no meaning to say that the ant is staying on some sort of a 2-dimensional surface, or the ant has magically escaped the 2-dimensional void and entered the

3-dimensional void. Not only is it meaningless to say the "void" is curved or flat, it is even meaningless to say the "void" is 2-dimensional, 3-dimensional or 4 dimensional, etc.

Of course, the ants can still find whether their space is flat or curved by measuring distances from inside. They may build a triangle with taught strings (Figure 10.8) and measure the sum of its three internal angles. However, this is also the only thing a human, as a superbeing can do. We 3-d humans are not superior to the 2-d ants. What is the geometry of "void" can only be found empirically. We 3-d humans even cannot "see" that it is curved. If the experiment finds that the sum of angles of a triangle (material model with taut strings) is different from π, we simply have a "different geometry" (different from Euclidean geometry) rather than a "curved space".

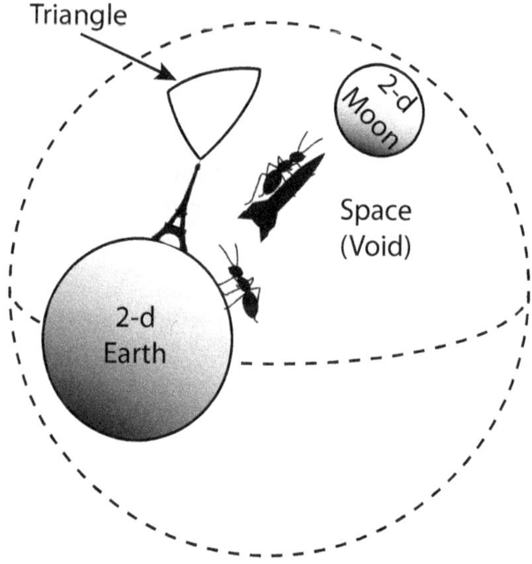

Figure 10.8   Material 2-dimensional curved space

The view of geometry in ancient Greece was mainly materialist. The figures they studied, like triangles and cubes, all have material models. It is funny that we discuss the Euclidean space every day, but Euclid never used the term "space" in his *Elements*. What we call "space geometry" today was called solid geometry, or geometry of solids then. The word for space in German is *Raum*, or literally, "room". At the time of Euclid, people were more concerned with the "furniture" in the "room", rather than the "room" itself, as the container. The word *geometry* literally means the measurement of the material earth, rather than measurement of the void space. Later in history, the concept of space has been abstracted from the concept of solids but went in the wrong direction. Geometry has become the study of the measurement of the void space. Newton even went to the extreme as to use the absolute space as a reference system to depict the motion of bodies.

Imagine in the far future we migrate to a remote galaxy. Is there a possibility in that galaxy that the geometry is Riemannian rather than Euclidean? The answer to the question is still subject to human conventions the same way we face on earth. It has all the possibilities: the platinum bar does not change length when moved to a different place and the geometry is Euclidean; the platinum bar changes its length when moved to a different place due to physics laws and the geometry is non-Euclidean (Riemannian); the platinum bar does not change length when moved to a different place and the geometry is non-Euclidean; the platinum bar changes its length when moved to a different place due to physics laws and the geometry is Euclidean.

What is the geometry of the universe? Is it flat or curved? Is it finite or infinite? There are no correct answers to these questions because these are wrong questions. Geometry is a human convention. Geometry and physics must work together to describe the world. We may choose a different geometry, provided that we adapt our physical laws to the new geometry, we then describe the same phenomena. Suppose the universe is bounded (meaning there exists a big constant D such that the distance d between any two objects $d < D$). We can still choose different geometries to describe it. We could choose a closed manifold with geometry of curved space like S^3, or we could choose Euclidean geometry in which the matter is held together in a bounded region by physical attractive forces while surrounded by infinite void. However, as void is nothingness, it even has no meaning to say the void is infinite or finite.

Abstract mathematics can be applied to reality, if and only if we find a physical model in reality for the mathematical structure. The n-dimensional Euclidean geometry can be applied, if we use a sequence of numbers $(x_1, \ldots, x_n)$ to represent a point in $\mathbb{E}^n$. This is a model within mathematics itself. Do we have a model of $\mathbb{E}^n$ in the real world? Yes, but not in the sense that we have 4-dimensional or higher dimensional solids that we can touch and feel. Think of a gray-scale digital image of $1,000 \times 1,000$ pixels. The image can be represented by one million numbers $(x_1, \ldots, x_{1,000,000})$. So each image is a point in a one-million-dimensional Euclidean space $\mathbb{E}^{1,000,000}$.

All the image processing applications, like image classification, recognition, segmentation, etc., are done in this Euclidean space. Even the processing of images in the human brain can be viewed as the processing in higher dimensional space. So we encounter higher dimensional spaces everyday, but not in the sense that we live in 4-dimensional space where we can touch and feel 4-dimensional solids or animals. Oftentimes, the images in our samples live in a lower dimensional subspace, or submanifold of $\mathbb{E}^n$. To find such lower dimensional subspace or submanifold is the essential task in machine learning—dimensionality reduction and manifold learning. This is not limited to images. In any machine learning applications, the data can be represented as points in the feature space, which is a higher dimensional Euclidean space, or tensor space (see Chap. 2).

§6. What Is General Relativity Exactly?

It is often a common understanding that the theory of general relativity applies Riemannian geometry, but there is a caveat. It is not exactly Riemannian geometry. The curved spacetime is a pseudo-Riemannian manifold, rather than a Riemannian manifold. This fact is not emphasized enough in many books on general relativity so that it causes misunderstandings.

It is a common understanding that the "curved" spacetime in general relativity is described by a manifold with a metric $g_{\mu\nu}$, which encodes all the information of spacetime properties. (One online tutorial explains that "the most important tensor in general relativity is the metric tensor $g_{\mu\nu}$, which helps us measure the lengths and angles in the curved geometry of spacetime." This is a misconception, because there are no lengths and angles in the spacetime. Namely, lengths and angles are not defined in spacetime.)

The essence of general relativity is often interpreted as follows: the gravity causes spacetime to curve; the curved spacetime governs matter how to move. In John Wheeler's words: "Spacetime tells matter how to move; matter tells spacetime how to curve." More precisely: the matter distribution (represented by the energy-momentum tensor $T_{\mu\nu}$) determines the spacetime metric $g_{\mu\nu}$ according to the Einstein's field equation:

$$R_{\mu\nu} - \tfrac{1}{2}Rg_{\mu\nu} = 8\pi GT_{\mu\nu}, \tag{10.17}$$

where $R_{\mu\nu}$ is the Ricci curvature tensor, R is the curvature scalar and G is Newton's gravitational constant (both $R_{\mu\nu}$ and R are determined by the metric $g_{\mu\nu}$). A mass particle (or light) moves along geodesic lines in the curved spacetime.

However, are the geodesics and the curvature of the spacetime in general relativity the same as the geodesics and curvature in Riemannian geometry? Let us first look at an outline of Gauss' theory of curved surfaces and Riemann's theory of curved spaces.

It was discovered by Gauss that all the intrinsic properties of a surface are determined by the differential quadratic form,

$$ds^2 = \sum_{i,j=1}^{2} g_{ij} dx_i dx_j.$$

For two points on a curve, if they are infinitely close to each other, the meaning of ds, or $\sqrt{ds^2}$, is the infinitesimal length of the segment of the curve between these two points. The square ds^2 of the line element ds is also called the metric. Geodesic lines are the curves that locally minimize the curve length between two points (see Sec. 1 for the meaning of Gaussian curvature).

Riemann generalized this to any dimension n, with

$$ds^2 = \sum_{i,j=1}^{n} g_{ij} dx_i dx_j. \tag{10.18}$$

Suppose we have a curve parameterized by t, and points A_1 and A_2 on the curve have parameters t_1 and t_2. The curve length between A_1 and A_2 is

$$L(\gamma) = \int_{A_1}^{A_2} ds = \int_{t_1}^{t_2} \mathscr{L} dt, \tag{10.19}$$

where

$$\mathscr{L} = \sqrt{\sum_{i,j=1}^{n} g_{ij} \dot{x}_i \dot{x}_j},$$

and $\dot{x}_i = dx_i/dt$. The geodesic line connecting A_1 and A_2 is a curve that locally (when A_1 and A_2 are close to each other) minimizes the arc length Eq. 10.19, with a necessary condition

$$\delta L(\gamma) = 0. \tag{10.20}$$

Another equivalent way of finding geodesics on a Riemannian manifold is to consider the "energy functional"

$$E(\gamma) = \int_{t_1}^{t_2} \mathscr{E} dt, \tag{10.21}$$

where

$$\mathscr{E} = \mathscr{L}^2 = \sum_{i,j=1}^{n} g_{ij}\dot{x}_i\dot{x}_j. \tag{10.22}$$

All minima of $E(\gamma)$ are also minima of $L(\gamma)$, but the set of minima for $L(\gamma)$ could be bigger since paths that are minima of $L(\gamma)$ can be arbitrarily re-parameterized (without changing their length), while minima of $E(\gamma)$ cannot.

Using calculus of variations, the necessary condition for $E(\gamma)$ to be minimum is

$$\delta E(\gamma) = 0. \tag{10.23}$$

This leads to the Euler-Lagrange equation

$$\frac{d}{dt}\frac{\partial\mathscr{E}}{\partial\dot{x}_k} - \frac{\partial\mathscr{E}}{\partial x_k} = 0,\ k = 1,\ldots,n. \tag{10.24}$$

Substituting $\mathscr{E}$ from Eq. 10.22 into Eq. 10.24, we obtain the geodesic equation

$$\frac{d^2 x_k}{dt^2} + \sum_{i,j=1}^{n} \Gamma_{ij}^{k}\frac{dx_i}{dt}\frac{dx_j}{dt} = 0, \tag{10.25}$$

where

$$\Gamma_{ij}^{k} = \frac{1}{2}\sum_{m=1}^{n} g^{km}\left(\partial_i g_{jm} + \partial_j g_{im} - \partial_m g_{ij}\right)$$

are the Christoffel symbols.

The Riemann curvature tensor (also called Riemann-Christoffel tensor) is

$$R_{ijk}^{l} = \partial_j\Gamma_{ik}^{l} - \partial_k\Gamma_{ij}^{l} + \sum_{m=1}^{n}(\Gamma_{ik}^{m}\Gamma_{mj}^{l} - \Gamma_{ij}^{m}\Gamma_{mk}^{l}). \tag{10.26}$$

It measures how much the local geometry deviates (second order effect) from that of Euclidean space $\mathbb{E}^n$.

Let U be an open set in the Riemannian manifold M with local coordinates $(x_1,\ldots,x_n)$ and let the metric tensor be g_{ij} as in Eq. 10.18. Suppose we make a coordinate change with a good guess

$$x_i' = x_i'(x_1,\ldots,x_n),\ i = 1,\ldots,n, \tag{10.27}$$

and we wish to make the metric tensor $g_{\mu\nu}$ in new coordinates x_i' taking the diagonal form

$$ds^2 = d(x_1')^2 + \ldots + d(x_n')^2. \tag{10.28}$$

This is not always possible, depending on the metric $g_{\mu\nu}$. The sufficient and necessary condition for this possibility is that all components R^l_{ijk} vanish. That is to say,

$$R^l_{ijk} \equiv 0, \tag{10.29}$$

in the entire open set U. This result is due to Elwin Christoffel. In such a case, we say M is locally flat, or locally Euclidean in U.

In his paper The Foundations of the General Theory of Relativity, Einstein [(1916)] basically repeated these procedures (Eqs. 10.18–10.26) for Riemannian geometry, and then switched to the context of spacetime, and continued to call ds the "line element" (Linienelement), "arc length" (Bogenläge), "arc distance" (Bogendistanz) and ds^2 the "square of the line element" (Quadrates des Linienelements).

The key difference is, in the context of Riemannian geometry, ds^2 is a quadratic form which is positive-definite, while in the context of spacetime, ds^2 is a quadratic form which is indefinite. Someone may say, it is not a big deal. Yes, it is. In the context of spacetime, the notation ds^2 itself is misleading. It does not mean the square of ds. (Note ds is not defined, and ds^2 acting on a vector may even yield a negative value. Especially, ds does not have a meaning of any length.) The symbol ds^2 should be treated as a single symbol, representing an indefinite quadratic form. To avoid confusion, it would be a better idea to use a symbol like Φ, rather than ds^2.

A Minkowski space is a vector space equipped with a metric which is a symmetric nondegenerate bilinear form. To distinguish it from the metric (positive-definite) for Euclidean spaces, we should call it the pseudo-metric, even if we use the same notation $\langle \mathbf{u}, \mathbf{v} \rangle$. Pseudo-Riemannian manifolds are locally modeled on Minkowski spaces while Riemannian manifolds are locally modeled on Euclidean spaces. In Minkowski spaces, as well as pseudo-Riemannian manifolds, we cannot talk about length, distance or angle. If $\langle \mathbf{u}, \mathbf{v} \rangle = 0$, we should say that the vectors $\mathbf{u}$ and $\mathbf{v}$ are pseudo-orthogonal to each other, to distinguish from the case of "orthogonal" in Euclidean spaces and Riemannian manifolds, because it has no meaning of being "perpendicular". Some author[6] even writes "the time axis is everywhere at right angles to the spatial extension" and "the time axis is not everywhere at right angles to the spatial dimensions", which are misleading. Some author[7] defines

$$\|\mathbf{v}\| \overset{\text{def}}{=} \sqrt{|\langle \mathbf{v}, \mathbf{v} \rangle|} \tag{10.30}$$

---

[6]Berenda, C. W. (1942). The Problem of the Rotating Disk, *Phys. Rev.*, **62**, 280–290.

[7]Gourgoulhon, É. (2013). *Special Relativity in General Frames: from Particles to Astrophysics*, English translation, (Springer-Verlag).

as the "norm" of the vector $\mathbf{v}$, which is also misleading, because this is very different from the ordinary norm of a normed vector space. In a Euclidean space, $\|\mathbf{v}\| \overset{\text{def}}{=} \sqrt{\langle \mathbf{v}, \mathbf{v}\rangle}$ is a true norm and it induces a topology, because $\langle \mathbf{v}, \mathbf{v}\rangle \geq 0$ is guaranteed. Pay attention to the absolute value symbol under the square root in Gourgoulhon's definition in Eq. 10.30. In a Minkowski space (or a pseudo-Riemannian manifold), $\langle \mathbf{v}, \mathbf{v}\rangle \geq 0$ does not hold. Taking the absolute value allows one to take the square root alright, but it is still not a form, and it does not induce a topology. To see the difference, we compare examples of a 2-dimensional Euclidean space (Figure 10.9a) and a Minkowski space (Figure 10.9b). Figure 10.9(a) (shaded region) represents $y^2 + x^2 < 1$ in the Euclidean space, while Figure 10.9(b) (shaded region) represents $t^2 - x^2 < 1$ in Minkowski space. (We use t to name the axis to remind us that it is like the case of time in relativity.) Figure 10.10(a) represents the "pseudo-norm" $\|\mathbf{v}\| = \sqrt{|\langle \mathbf{v}, \mathbf{v}\rangle|} = \sqrt{|t^2 - x^2|} < 1$. Figure 10.10(b) represents the same condition $\|\mathbf{v}\| < 1$ but restricted to the time-like part of the spacetime, where $\langle \mathbf{v}, \mathbf{v}\rangle > 0$ and we can take the square root of it without first taking the absolute value. (We have adopted the convention of signature $(+---)$ for the spacetime.) Hence we see a big difference between the Minkowski space for spacetime and the Euclidean space.

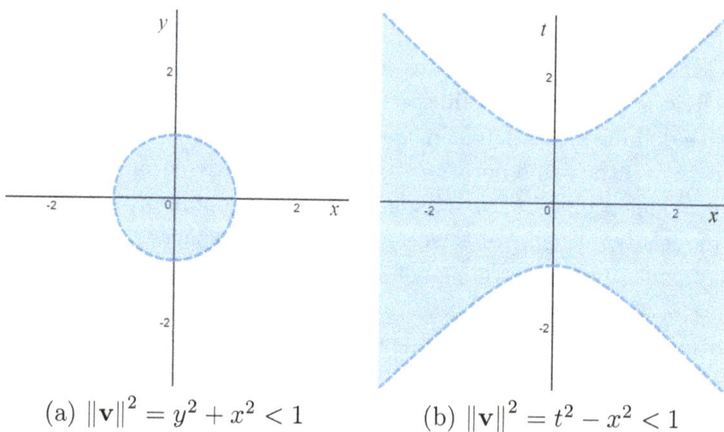

(a) $\|\mathbf{v}\|^2 = y^2 + x^2 < 1$ (b) $\|\mathbf{v}\|^2 = t^2 - x^2 < 1$

Figure 10.9   (a) Euclidean space (b) Minkowski space

Yes, even in the case of pseudo-Riemannian manifold, we may still mimic Riemannian geometry and find the necessary condition for the integral in Eq. 10.21 to have an extremal value, which is the same as Eq. 10.25. That

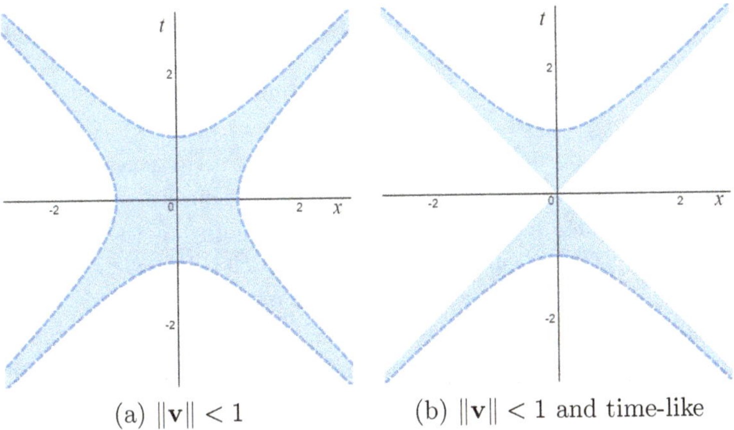

(a) $\|\mathbf{v}\| < 1$ (b) $\|\mathbf{v}\| < 1$ and time-like

Figure 10.10   Pseudo-norm for Minkowski space

was what Einstein did in 1916, but these no longer have meanings in the context of Riemannian geometry. Einstein still calls Eq. 10.25 "geodesic lines", but ds does not have a meaning of length, and these "geodesic lines" do not have "length minimizing" meaning any more. In general relativity, a mass particle is supposed to move along a "time-like" curve ($ds^2 > 0$). When restricted to time-like curves, we may take the square root of ds^2 and define $ds = \sqrt{(ds^2)}$, but ds corresponds to "proper time" rather than any length or distance. When restricted to time-like curves, the integral $\int ds$ on the "geodesic" described by Eq. 10.25 indeed takes an extremal value. However, it is the maximum, instead of the minimum. Eq. 10.26 is still called the "Riemann-Christoffel tensor" (the "curvature tensor") by Einstein, but it no longer has any metric meaning (compare with Figure 10.2 in Sec. 1). One result of Christoffel for Riemannian manifolds holds similarly for pseudo-Riemannian manifolds. That is, $R^l_{ijk} \equiv 0$ in an open set U is the sufficient and necessary condition for the existence of a coordinate transformation in the form of Eq. 10.27 such that the metric can be written in the form of

$$ds^2 = \pm d(x'_1)^2 \pm \ldots \pm d(x'_n)^2.$$

Namely, locally in the open set U, it is a flat, or pseudo-Euclidean space.

In the modern language, geodesics and curvature for pseudo-Riemannian manifolds have been defined. These are actually generalized and defined for an even more general class of mathematical entities—affine manifolds, or differentiable manifolds with an affine connection. These concepts are generalized in the affine sense. To make the generalization, let us first take another look of geodesics on Riemannian manifolds, especially on surfaces.

The geodesics on Riemannian manifolds may be looked at from two perspectives: in the metric sense and the affine sense. In the metric sense, they are the "shortest" lines. In the affine sense, they are the "straightest" lines. What do "straight" and "straightest" mean then?

Straight lines do not exist on a curved surface (in general) but we can look for the "straightest curves". The "straightest curves" are the geodesics on a curved surface and they are the best analogy to the straight lines. Euclid actually has given us some insight. Euclid defines "A straight line is a line which lies evenly with the points on itself." This is vague as a definition, but its insight is helpful.

Imagine we are driving a car on a plane. What makes the difference between a circle and a straight line? If we keep driving forward without turning left or right, we are going on a straight line (Figure 10.11a). That is, we lock the steering wheel to the forward position (with all four tires aligned perpendicular to the axles, or even better yet, we may manufacture a car without a steering wheel but with tires so aligned). To drive on the circle (Figure 10.11b), we must turn the steering wheel to the right. This reflects our intuition about the concept of straightness better. If we ask for directions on the street and someone tells us to "go straight", we understand it as "do not turn left or right", and we hardly think immediately to find a route to minimize the distance. It seems that the metric property—distance minimizing nature of the straight line—is only a coincidental fact that accompanies the straight lines. In terms of geometry, for a general curve, the tangent vectors of the curve keep changing directions, while the tangent vectors of a straight line do not change directions. In other words, tangent vectors of a straight line are parallel, while the tangent vectors of a curve are not parallel.

Now suppose we drive on a curved surface (like the earth, which is a sphere). If we keep the steering wheel straight, we trace a geodesic line (Figure 10.11c). So a geodesic line can be defined as a curve whose tangent vectors are parallel to each other. Geodesics defined this way are called affine geodesics. If we keep turning the steering wheel, then we trace a small circle (Figure 10.11d).

When we say the tangent vectors of a geodesic line are parallel to each other on a curved surface like a sphere, we must fix some details. In Figure 10.11(c), the tangent vectors $\mathbf{v}_1, \mathbf{v}_2, \mathbf{v}_3, \mathbf{v}_4$ are not parallel in the ambient Euclidean space. We must define this carefully—tangent vector $\mathbf{v}_1$ is parallel to $\mathbf{v}_2$ when transported from one location to a nearby location along a curve. This is known as Levi-Civita parallel transport and can be

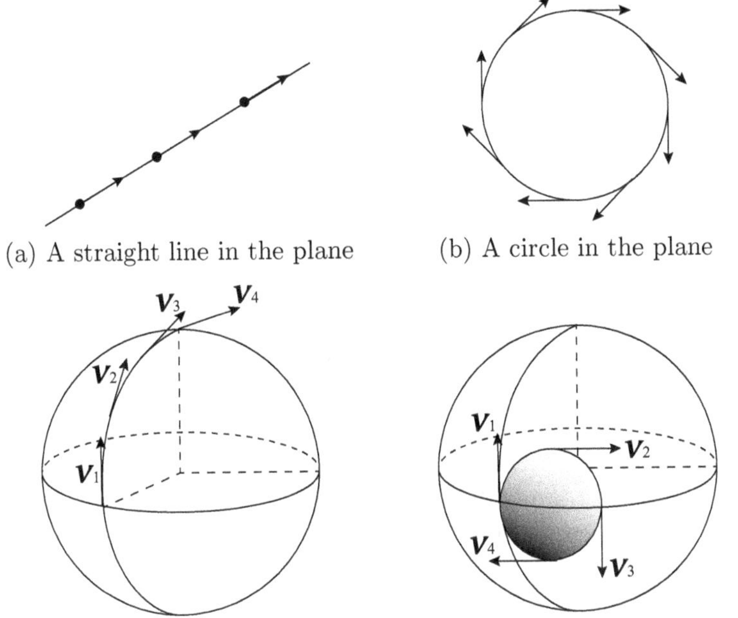

(a) A straight line in the plane   (b) A circle in the plane

(c) A geodesic line on the sphere   (d) A small circle on the sphere

Figure 10.11   Affine geodesics

defined with the help of the concept of distance (or metric) on a Riemannian manifold. The parallel transport determines an affine connection and vice versa. The Levi-Civita connection or parallel transport can be defined either operationally or axiomatically using the metric for a Riemannian manifold.

For any differentiable manifold (without a Riemannian metric), we generalize the affine connection via a set of axioms. A differentiable manifold with an affine connection, also called an affine manifold, is defined below.

Definition 1. Let M be a differentiable manifold, $\mathfrak{X}(M)$ be the set of all smooth vector fields, and $\mathcal{F}(M)$ be the set of all smooth scalar fields on M. M together with a mapping $\nabla : \mathfrak{X}(M) \times \mathfrak{X}(M) \to \mathfrak{X}(M)$; $(X, Y) \mapsto \nabla_X Y$, is called an **affine manifold**, or **affinely-connected manifold**, if ∇ satisfies the following axioms, for all smooth vector fields $X, X_1, X_2, Y, Y_1, Y_2 \in \mathfrak{X}(M)$ and smooth scalar fields $f \in \mathcal{F}(M)$.

(1) $\mathcal{F}(M)$-linear on the first variable:

(1a) $\nabla_{X_1+X_2} Y = \nabla_{X_1} Y + \nabla_{X_2} Y$;
(1b) $\nabla_{fX} Y = f\nabla_X Y$.
(2) Additive on the second variable:
$$\nabla_X(Y_1 + Y_2) = \nabla_X Y_1 + \nabla_X Y_2.$$
(3) Leibniz rule on the second variable:
$$\nabla_X(fY) = f\nabla_X Y + (\nabla_X f)Y.$$

$\nabla_X Y$ is called the **covariant derivative** of Y in the direction of X. ∇ is called an **affine connection** on M.

Note that $\nabla_X Y$ is $\mathbb{R}$-linear on both X and Y. It is $\mathcal{F}(M)$-linear on X, but not $\mathcal{F}(M)$-linear on Y. Therefore it is not a tensor field. It is not a tensor field because it is not defined as a tensor field, and we do not need to further test it with transformation laws. This is the modern view. In the old-fashioned component approach, it is explained that although the affine connection coefficients (e.g., Christoffel symbols) have three indices but they do not transform like of tensor (field) of degree 3.

For a vector field X on a curve γ, if the covariant derivative of X along curve γ (the tangent vector field of γ) is zero, then we say the vector field X on γ is parallel, or the vectors in X are parallelly transported along γ. An affine geodesic line is defined to be a curve whose tangent vectors are parallel with respect to the affine connection ∇. In other words, if γ is an affine geodesic line, its tangent vector remains tangent to γ after parallelly transported along γ itself (with respect to affine connection ∇).

The theory of affine connections was studied by H. Weyl [(1918), (1950)]. A word of caution here: many books simply treat Definition 1 as the definition of an affine connection, but it is more important to see this from the structural point of view (see Appendix 2). Definition 1 is a set of axioms for the affine manifold. We are introducing a new structure here. Namely, we are granting ourselves the new freedom of endowing an *arbitrary* connection to the differentiable manifold. The same differentiable manifold equipped with different affine connections will have very different properties. This generalization is a big step and it can be arbitrary and wild with new and unfamiliar examples.

Levi-Civita connection for the Riemannian manifold is induced by the metric. For an arbitrary affine connection, sometimes it is possible to find a metric such that the metric can induce the given affine connection, but there also exist non-metricizable affine connections. A. Vanžurová and P. Žáčková [(2009)] have given an example of a non-metricizable affine connection for a 2-dimensional differentiable manifold with the following components in local coordinates:

$$\Gamma^1_{11} = \Gamma^1_{22} = 1, \ \Gamma^1_{12} = \Gamma^1_{21} = 0, \ \Gamma^2_{11} = \Gamma^2_{22} = 0, \ \Gamma^2_{12} = \Gamma^2_{21} = 2.$$

We can calculate the Ricci curvature tensor $R_{11} = -2$, $R_{12} = R_{21} = 0$, $R_{22} = -1$, and the Ricci scalar curvature $R = -3$.

Remark. Affine manifolds are the generalizations of Riemannian manifolds. Let us look at their similarities and differences. Parallel transport is an important concept for both of them.

(1) Parallel transport and covariant derivative can define each other. If covariant derivative is defined first, we can define parallel transport as this: a vector field is parallel if its covariant derivative is zero. If parallel transport is defined first, we can define the covariant derivative of a vector field $\mathbf{v}(p)$ at a point p as: go to a nearby point p', parallel transport the vector $\mathbf{v}(p')$ back to p and it becomes $\mathbf{v}'(p)$. Find the difference $\mathbf{v}'(p) - \mathbf{v}(p)$, and find out the change rate in the limit process when p and p' are infinitely close (in the sense of topology and not necessarily in distance).

(2) Using parallel transport (or covariant derivative), we can define the Riemann curvature tensor.

(3) Differences: for Riemannian manifold, we can define the geodesics first using the metric property (locally length minimizing Eqs. 10.23, 10.24 and 10.25. We can define Levi-Civita parallel transport operationally using the concept of geodesics. The tangent vectors to a geodesic line are parallel by definition, if their lengths are equal. We can first define parallel transport along a geodesic line in a 2-dimensional Riemannian manifold: the point of origin of the vector moves along the geodesic, and the vector itself moves continuously so that its angle with the geodesic and its length remain constant. To define parallel transport along an arbitrary curve, we approximate the curve by a broken line consisting of geodesic arcs, and take the limit for the length of each geodesic arc tending to zero. For higher dimensions, more details need to be fixed, but we omit that case here. See more in [Arnold (1997) pp. 301–306] and [Guo (2014) pp. 313–325]. For the generalized affine manifolds, the order is rather opposite. We first stipulate what is parallel transport (or covariant derivative, or affine connection) abstractly and arbitrarily (loosely constrained by the three axioms in Definition 1). Then a geodesic line is defined as a curve whose tangent vectors are parallel with respect to the stipulated rules of parallel transport (or affine connection). For Riemannian manifolds, the Levi-Civita parallel

transport can also be defined axiomatically. That is, the Levi-Civita connection is the only torsion-free affine connection that satisfies the axioms in Definition 1 and also preserves the Riemannian metric (the lengths and angles of vectors). The operational definition is more intuitive pedagogically. Similarly we can define an affine connection for a pseudo-Riemannian manifold as the unique torsion-free affine connection that satisfies the axioms in Definition 1 and also preserves the pseudo-Riemannian metric, but the operational definition for Riemannian manifolds does not carry over to the pseudo-Riemannian manifolds, because the geodesic line needs to be defined using parallel transport (or affine connection) as a prior concept. What is similarly true for a pseudo-Riemannian manifold is that if two vectors $\mathbf{u}$ and $\mathbf{v}$ at the same point are parallelly transported to another point along a curve, $\langle \mathbf{u}, \mathbf{u} \rangle$, $\langle \mathbf{v}, \mathbf{v} \rangle$ and $\langle \mathbf{u}, \mathbf{v} \rangle$ are all invariant, except these do not have interpretations of lengths and angles because the inner product $\langle \cdot, \cdot \rangle$ now is no longer positive-definite.

*** Methodology**: Generalization

It is true that through the definition of the abstract affine connection, we can define affine geodesics as the "straightest" curves in a sense. However, we must bear in mind that the affine connection so defined is a *generalization* of the Levi-Civita connection, which we are familiar with. After the generalization, some properties are kept while others are lost. The affine connection defined by these axioms abstractly can be arbitrary and wild. The geodesics and the sense of "straightness" so defined are the generalized concepts and can be arbitrary and wild. They may not bear much similarity (other than satisfying those axioms) to our ordinary concept of straightness (like in Figure 10.11d) on surfaces or Riemannian manifolds. The following is another example to illustrate how different a concept can be after being generalized.

In the Euclidean space, the Euclidean distance between two points is defined as

$$d(p, q) \overset{\text{def}}{=} \sqrt{(p_1 - q_1)^2 + \ldots + (p_n - q_n)^2}.$$

The concept of distance can be generalized by the following axioms: for all p, q, s,

(1) $d(p, q) = 0$ if and only if $p = q$.

(2) $d(p, q) = d(q, p)$ (symmetry).

(3) $d(p, q) + d(q, s) \geq d(p, s)$ (triangle inequality).

A nonempty set together with such a distance defined is called a metric space, which is the generalization of Euclidean spaces. The generalized metric spaces can be very different from the Euclidean spaces, even if we still call $d(p, q)$ the distance. For example, we may define

$$d(p, q) \overset{\text{def}}{=} \begin{cases} 1, & \text{if } p \neq q, \\ 0, & \text{if } p = q. \end{cases}$$

This is called the discrete metric.

Many known properties and theorems for Euclidean spaces will fail in this discrete metric space. We list a few strange properties for this discrete metric space:

(1) Every subset is an open set and a closed set.

(2) A sequence is convergent only if it is a constant sequence after finite initial terms, $\ldots, x, x, x, \ldots$.

(3) Every function is a continuous function.

There is another caveat. When we say a the motion of a particle (or light) is along a geodesic, it means an affine geodesic in the *4-dimensional spacetime*. In a local neighborhood, we may define a 3-dimensional submanifold of the spacetime manifold—the space manifold at simultaneous time t. The pseudo-metric of the spacetime induces a positive-definite metric on this space manifold and makes it a Riemannian submanifold. However, in general the trajectory of the motion of a mass particle (or light) is not a geodesic line in this space manifold with respect to the induced Riemannian metric. In short, it is not a geodesic line in the 3-d "space".

Shakespeare writes in *Romeo and Juliet*, "What's in a name? That which we call a rose by any other name would smell as sweet."

If different names are used to distinguish different things, and the generalized affine geodesic is renamed *cisedoeg* ("geodesic" spelled backward), the generalized affine curvature is renamed *erutavruc* ("curvature" spelled backward), general relativity would be interpreted as "gravity changes the spacetime *erutavruc* and particles move along *cisedoegs*". Would the beauty of general relativity be appreciated the same?

In his papers General Relativity and Flat Space (I and II), N. Rosen [(1940)] points out, general relativity can be described in flat Minkowski space. Suppose we use the same coordinates as in Eq. 10.18 but dictate a different pseudo-metric

$$d\sigma^2 = \sum_{\mu\nu} \gamma_{\mu\nu} dx_\mu dx_\nu,$$

where $\gamma_{\mu\nu}$ are constants, so that the spacetime is flat. The equation of motion Eq. 10.25 can be translated to a different form. In fact, Rosen is making a transformation from the Einstein's manifold to a flat Minkowski space. Of course, this transformation is diffeomorphic but not pseudo-isometric, meaning it does not preserve the quadratic form ds^2. This idea is similar to the Mercator metric we have used for the surface of the earth in Sec. 5, which results in a flat plane Euclidean geometry. How did we obtain this flat Mercator metric? It is pretty simple. We wrap the earth (sphere) tightly with a cylinder and project the sphere onto this cylinder (Figure 10.12). The geometry on the cylinder is of course flat. Think of that the globe is a transparent model and we put a light bulb at the center of the globe. Some characters like Mickey mouse are performing a show on the globe. These characters will cast shadows on the wall of the cylinder.[8] We can either choose to watch these characters directly on the globe, or watch the shadows on the cylinder.

 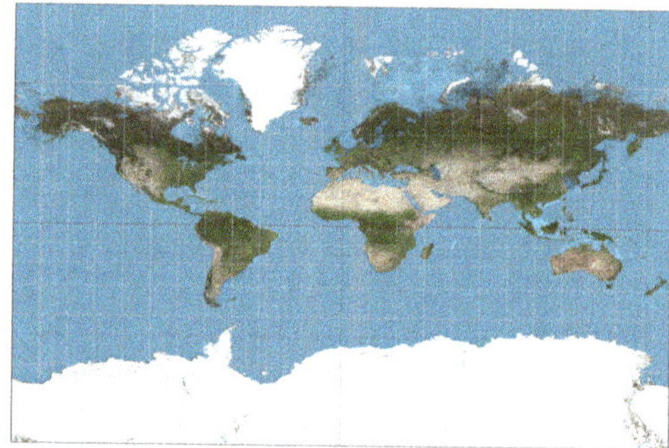

Figure 10.12   Mercator cylindrical projection

[8]Mercator projection is one of the cylindrical projections, in which meridians are projected to the vertical lines of the wrapping cylinder. However, the north-south stretch of the meridian in different cylindrical projections can be an arbitrary smooth function, not necessarily the perspective projection, or any simple geometric projection. The north-south stretch Eq. 10.16 in Mercator projection is not perspective, but this is not essential to explain how the sphere is projected onto a plane using cylindrical projections.

What game is Rosen playing? In my opinion, spacetime metrics $g_{\mu\nu}$ and $\gamma_{\mu\nu}$ represent different conventions of length and time measurement standards. Rosen is suggesting a set of length and time measurement standards which is different from what Einstein suggested.

My interpretation of general relativity is: at each point in spacetime, if we adopt length and time measurement standards $g_{\mu\nu}$ in accordance with Eq. 10.17, then a mass particle or light travels along a *cisedoeg* line (in the 4-dimensional spacetime). Gravity does not curve spacetime. Man curves it. In fact, in general relativity, $g_{\mu\nu}$ represents the human manipulation of the length and time measurement standards. Do not take for granted that Einstein's pseudo-metric $g_{\mu\nu}$ for the curved spacetime is the true description of the show while Rosen's $\gamma_{\mu\nu}$ for the flat spacetime is the shadow. It could be just the other way around. The difference is only a matter of convenience.

Poincaré [(1905)] remarked: "If Lobatschewsky's geometry is true, the parallax of a very distant star will be finite. If Riemann's is true, it will be negative. These are the results which seem within the reach of experiment, and it is hoped that astronomical observations may enable us to decide between the three geometries. But what we call a straight line in astronomy is simply the path of a ray of light. If, therefore, we were to discover negative parallaxes, or to prove that all parallaxes are higher than a certain limit, we should have a choice between two conclusions: we could give up Euclidean geometry, or modify the laws of optics, and suppose that light is not rigorously propagated in a straight line."

§7. What Is Time Exactly?

In a recent paper [Guo (2021)],[9] I argued further for this conventionalist view. A common belief today is that the Lorentz transformation is correct but the Galilean transformation is wrong (only approximately correct in low speed limit). However, in general relativity [Einstein (1916)], any form of smooth coordinate transformation

$$\begin{aligned}
x' &= x'(x, y, z, t), \\
y' &= y'(x, y, z, t), \\
z' &= z'(x, y, z, t), \\
t' &= t'(x, y, z, t),
\end{aligned} \tag{10.31}$$

is equally valid.

---

[9]Guo, H. (2021). A New Paradox and the Reconciliation of Lorentz and Galilean Transformations, *Synthese*, https://doi.org/10.1007/s11229-021-03155-y (open access).

The Galilean transformation

$$x' = x - vt,$$
$$y' = y,$$
$$z' = z,$$
$$t' = t,$$

$$(10.32)$$

is certainly just a special case of Eq. 10.31, and should be as valid as the Lorentz transformation. This renders a new paradox [Guo (2021)] (see Figure 10.13).

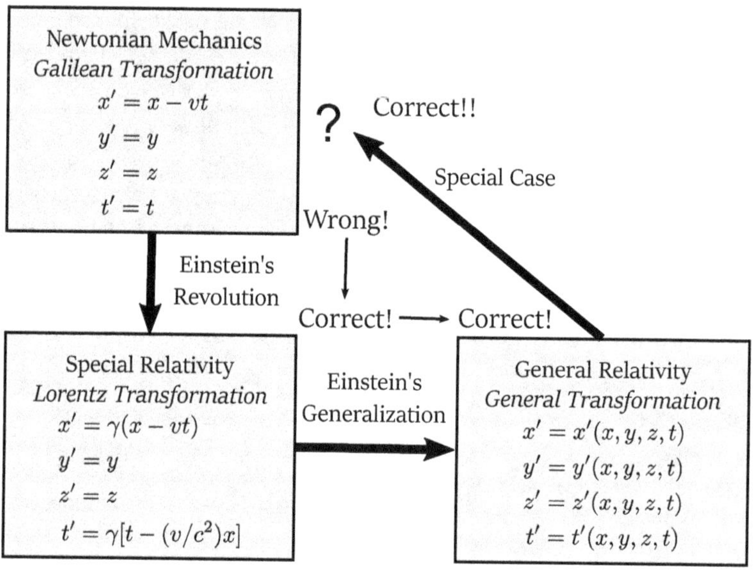

Figure 10.13   My paradox

*** Misconception**: Galilean transformation implies infinite (light) signal speed.

First let us clarify a concept in logic. Look at the propositions in the following:
(A): If p, then q. In symbolic language, $p \rightarrow q$.
(B): If q, then p. In symbolic language, $q \rightarrow p$.

Proposition B is called the converse of proposition A. If A is true, B may or may not be true. If B is true, A may or may not be true.

Now we apply this to the Galilean transformation. Suppose the following:

p: The coordinates in frames K and K$'$ are related by the Galilean transformation.

q: There exist (light) signals with infinite speed.

Now the two propositions become:

(A): Galilean transformation implies the existence of infinite (light) signal speed.

(B): If there exist (light) signals with infinite speed, then we can implement time and coordinates in frames K and K$'$ such that they are related by the Galilean transformation.

First we show that B is true:

Let us assume that the light speed is infinite. When the clock at location A reads time t_A, we send a light signal to location B with infinite speed, and set the clock at B the same as t_A. The clocks at different locations coordinated in such a way will obey the Galilean transformation.

With the education in logic, we understand that B is true does not imply its converse A is true. In fact, in this case, A is false, which is a commonly held misconception. We shall show why A is false in the following. That means, without using infinite speed signals, we are still able to implement time and coordinates in frames K and K$'$ such that they are related by the Galilean transformation.

This misconception sometimes appears in a different form as follows.

*** Misconception**: Galilean transformation is wrong because there is not a practical operational synchronization procedure to implement the time represented by this transformation.

Proposition 1. *The Galilean transformation and Lorentz transformation represent two conventions of time and length measurement in reference frame K$'$. They are equivalent to each other when describing physical phenomena.*

In the following, we shall demonstrate a practical operational synchronization procedure to implement the Galilean transformation, given finite speed of light signals.

Let K be an inertial frame with coordinates (x, y, z, t), in which we adopt the Einstein convention such that the light speed is isotropic with a constant c. Let us call K the *primary reference frame.* Suppose reference frame K$'$ is moving at velocity v with respect to K along x-direction, and (x'_E, y'_E, z'_E, t'_E) are the Einstein coordinates in reference frame K$'$. The Einstein coordinates in reference frames K and K$'$ are related by the Lorentz transformation

$$
\begin{aligned}
x'_E &= \gamma\left(x - vt\right), \\
y'_E &= y, \\
z'_E &= z, \\
t'_E &= \gamma\left(t - \frac{v}{c^2}x\right),
\end{aligned}
\tag{10.33}
$$

where $\gamma = 1/\sqrt{1 - v^2/c^2}$. Now we make some adjustment of coordinates and time in frame K$'$. We define a coordinate transformation *inside* reference frame K$'$ as follows:

$$
\begin{aligned}
x'_N &\overset{\text{def}}{=} \frac{1}{\gamma}x'_E, \\
y'_N &\overset{\text{def}}{=} y'_E, \\
z'_N &\overset{\text{def}}{=} z'_E, \\
t'_N &\overset{\text{def}}{=} \gamma t'_E + \frac{\gamma v}{c^2}x'_E,
\end{aligned}
\tag{10.34}
$$

This is also a special case of the general transformation Eq. 10.31 and is allowed by Einstein in general relativity. We call x'_N, y'_N, z'_N the Newton coordinates and t'_N the N-time. We call x'_E the E-time. Straightforward calculation reveals that

$$
\begin{aligned}
x'_N &= x - vt, \\
y'_N &= y, \\
z'_N &= z, \\
t'_N &= t.
\end{aligned}
\tag{10.35}
$$

This is just Galilean transformation (see Figure 10.14).

The inverse transformation of Eq. 10.34 can be found to be

$$x'_E = \gamma x'_N,$$
$$y'_E = y'_N,$$
$$z'_E = z'_N,$$
$$t'_E = \frac{1}{\gamma} t'_N - \frac{\gamma v}{c^2} x'_N,$$

(10.36)

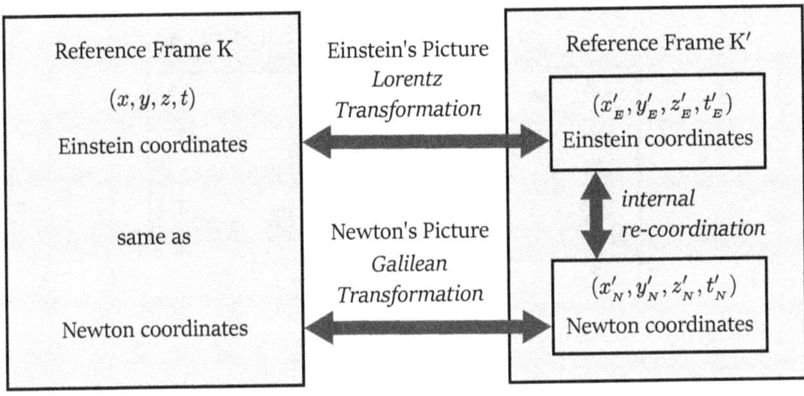

Figure 10.14   Equivalence of Galilean transformation and Lorentz transformation

This can be put in an analogy in quantum mechanics. It is similar to the relationship between the Schrödinger's picture and Heisenberg's picture. In the Schrödinger's picture, the operators representing physical observables are static while the quantum states evolve with time. Historically, Heisenberg took a very different approach. In the Heisenberg's picture, the state is static while the operators (or infinite dimensional matrices) representing physical observables evolve with time. If we just look at the appearances of these two theories, they seem to be totally different and unrelated. It was Dirac who revealed their relationship. They are related by a basis change in the Hilbert space. Hence Dirac united the two theories in the framework of the abstract Hilbert space. The Schrödinger's picture and the Heisenberg's picture are equivalent, and they describe the same physical phenomena (Figure 10.15).

Eq. 10.34 suggests a practical method to implement a system of clocks called N-clocks for reference frame K', which tell the N-time t'_N. At each location in space we place an N-clock side by side with the E-clock. The

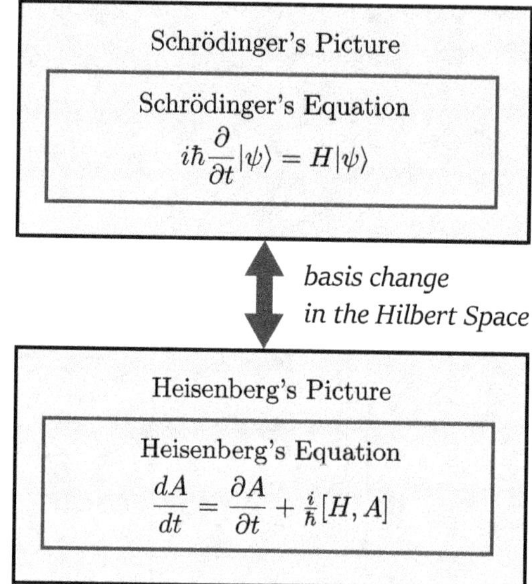

Figure 10.15   Analogy in quantum mechanics: equivalence of Schrödinger's picture and Heisenberg's picture

N-clock can be implemented using an E-clock by embedding a computer chip in it. The computer takes the E-time t'_E and its own coordinate x'_E (and the parameter v) as input and then computes its output t'_N according to Eq. 10.34. This should not be a novel idea, nor difficult in practice, as all the modern atomic clocks have utilized sophisticated electronic circuits in them.

There might be an objection to this idea of implementing N-clocks: the computation on the computer may take a time delay, rather than giving an instant output from the input. In fact, the implementation of N-clocks can even be achieved in a much easier way, without using a computer doing the translation on the fly all the time. We realize that at any particular location with coordinate x'_E in K', t'_N is linearly related to t'_E in such a way $t'_N = \gamma t'_E + a$, where γ is a scaling factor depending on the speed v only, and $a = (\gamma v/c^2)x'_E$ is an offset, which is a constant at each location x'_E. We only need to re-calibrate the E-clock to obtain an N-clock by re-labeling the time unit on the clock by a factor of $1/\gamma$ and then adding a constant a once and for all.

Note that now we have two sets of coordinate and time systems for the *same reference frame* K'. E-time is no longer the unique God-given time standard for K′. When we speak of time, we must make clear whether it is E-time or N-time to avoid confusion, both of which are equally legitimate. Newton time coordination provides a different simultaneity standard from Einstein simultaneity. When we talk about distance, we must make clear whether it is E-distance or N-distance. When we talk about speed, we must make clear what coordinates and time we are using. If we use E-distance and E-time, we get E-speed. If we use N-distance and N-time, we get N-speed. Hence the N-speed of light is not a constant in K′, but this is just a different description. Both describe the same physical phenomena.

This formulation might seem the comeback of Newton's absolute space and absolute time, and the primary frame K might look like the absolute ether reference frame, but this is not the case. The primary frame K is an arbitrarily choice by convention. Any inertial frame can be chosen as the primary frame K. It is "preferred by humans", but not "privileged by nature".

In reference frame K′ we define the E-velocity to be

$$\mathbf{u}_E \stackrel{\text{def}}{=} (u_{xE}, u_{yE}, u_{zE}) \stackrel{\text{def}}{=} \left(\frac{dx'_E}{dt'_E}, \frac{dy'_E}{dt'_E}, \frac{dz'_E}{dt'_E} \right), \tag{10.37}$$

and N-velocity to be

$$\mathbf{u}_N \stackrel{\text{def}}{=} (u_{xN}, u_{yN}, u_{zN}) \stackrel{\text{def}}{=} \left(\frac{dx'_N}{dt'_N}, \frac{dy'_N}{dt'_N}, \frac{dz'_N}{dt'_N} \right). \tag{10.38}$$

It is straightforward to find that E-velocity and N-velocity are related by

$$\begin{aligned}
u_{xN} &= \frac{u_{xE}}{\gamma^2 \left(1 + \frac{vu_{xE}}{c^2}\right)}, \\
u_{yN} &= \frac{u_{yE}}{\gamma \left(1 + \frac{vu_{xE}}{c^2}\right)}, \\
u_{zN} &= \frac{u_{zE}}{\gamma \left(1 + \frac{vu_{xE}}{c^2}\right)},
\end{aligned} \tag{10.39}$$

and its inverse transformation is

$$\begin{aligned}
u_{xE} &= \frac{u_{xN}}{1 - \frac{v(v+u_{xN})}{c^2}}, \\
u_{yE} &= \frac{u_{yN}}{\gamma \left(1 - \frac{v(v+u_{xN})}{c^2}\right)}, \\
u_{zE} &= \frac{u_{zN}}{\gamma \left(1 - \frac{v(v+u_{xN})}{c^2}\right)}.
\end{aligned} \tag{10.40}$$

Let us look at a few examples. For a light beam with E-velocity $u_{x_E} = c$, it translates to N-velocity

$$u_{x_N} = c - v. \tag{10.41}$$

For a light beam with E-velocity $u_{x_E} = -c$, it translates to N-velocity

$$u_{x_N} = -(c + v). \tag{10.42}$$

Take another example. Let $v = 0.8c$. A mass particle with E-velocity $u_{x_E} = 0.9c$ translates to N-velocity $u_{x_N} = 0.188c$. E-velocity $u_{x_E} = -0.9c$ translates to N-velocity $u_{x_N} = -1.16c$. The magnitude of N-velocity can exceed c but the physics is the same. Newton coordination is completely as valid as Einstein coordination. Any physical phenomena which can be described by Einstein coordination can be described by (or translated to) Newton coordination as well.

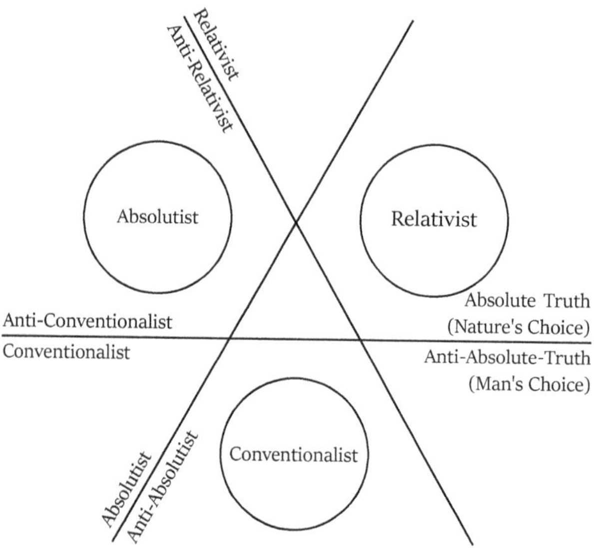

Figure 10.16   Three schools in philosophy

The topics we have discussed above involve a key issue in the theory of relativity—conventionality, which is still an unsettled debate today. Regarding time and simultaneity, there are three schools of opinions: the absolutist, the relativist, and the conventionalist. Each school is opposed to the *other two schools*. Each school believes the other two schools are wrong (see Figure 10.16).

The absolutists are represented by Issac Newton. In modern days people holding the absolutist view are very few. These include Albert Michelson, Georges Sagnac, Herbert Ives, Herbert Dingle, Franco Selleri, Louis Essen, Paul Marmet and Petr Beckmann. The mainstream has been the strong relativist/anti-conventionalist view. Research papers supporting the conventionalist view are very difficult to pass the journal peer review systems today. Einstein had a limited conventionalist view [Einstein (1961), p. 25]. Some authors hold a strong anti-conventionalist view [Ohanian (2009); Friedman (1983); Malament (1977)]. Ohanian published a book entitled *Einstein's Mistakes* [Ohanian (2009)]. A number of the "mistakes" of Einstein that Ohanian criticized in this book are actually the conventionalist views of Einstein, which in my opinion are not mistakes.

F. Selleri holds an absolutist view. He believes that the synchronization in a reference system is not conventional but rather the nature's choice. He believes the "absolute simultaneity", meaning if two events are simultaneous in one inertial frame then they are simultaneous in any other inertial reference frame. He considers both Lorentz and Galilean transformations are wrong, and the only correct coordinate transformation is the so-called "inertial transformation" [Selleri (1996)]. He has raised a paradox [Selleri (1997)]. He starts with the argument: on the circle of the rim of a rotating disk, the light speed is anisotropic. He considers the limit when the radius of the disk $r \to \infty$, but the angular velocity of the disk $\omega \to 0$ while keeping $v = \omega r$ constant. In this limit, the acceleration on the rim tends to zero. Therefore locally on the rim, it is effectively an inertial frame, but the anisotropic light speed survives this limit process and contradicts Einstein's theory of special relativity. In Selleri's words: "we must conclude that the famous synchronisation problem is solved by nature itself: it is not true that the synchronisation procedure can be chosen freely because Einstein's convention leads to an unacceptable discontinuity in the physical theory."

My argument that the Galilean transformation is equivalent to Lorentz transformation provides a straightforward resolution of the Selleri's paradox [Guo (2021)] (see Figure 10.17).

Some scholars may agree on that the Newton coordinates x'_N, y'_N, z'_N, t'_N for reference frame K′ with Galilean transformation are valid, but they argue that these coordinates are not *physical*. The only *physical* coordinates for K′ are the Einstein coordinates x'_E, y'_E, z'_E, t'_E. I disagree. The term *physical* is very vague and it has a connotation of "the unique choice" by nature, or by God. The *physical time* is understood the same as the *true*

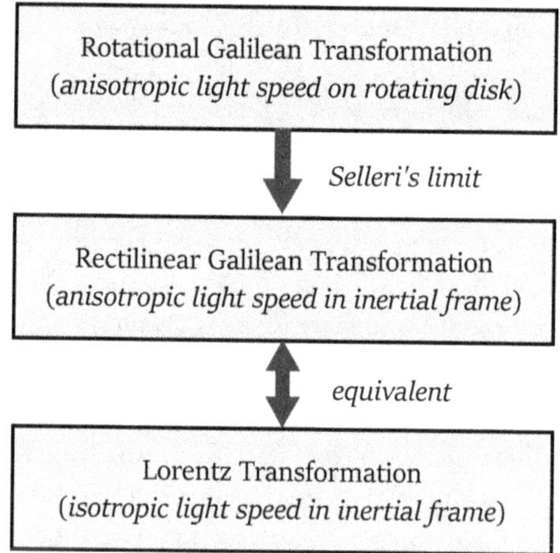

Figure 10.17   Resolution of Selleri's paradox

time. However, people disagree on what is physical, because everyone is making his own choice on behalf of God. Newton would certainly think that the absolute space, absolute time and Galilean transformation are physical. Selleri thinks the "inertial transformation" is the "nature's choice" and it must be physical [Selleri (1997)]. Lorentz introduced the transformation with coordinates x'_E, y'_E, z'_E, t'_E now bearing his name, prior to Einstein. To Lorentz, the Newton time t'_N was the true time or physical time. He timidly and humbly called t'_E the "local time", meaning it was just an intermediate, or temporary variable without the meaning of true time. Einstein then switched the course by promoting Lorentz's "local time" t'_E to the status of the true time or physical time. Since then the Newton time t'_N has been demoted. A new dynasty has overthrown and replaced the old dynasty. However, Einstein launched his revolution against Newtonian time using exactly a conventionalist argument [Einstein (1961)]: "That light requires the same time to traverse the path $A \to M$ as for the path $B \to M$ is in reality *neither a supposition nor a hypothesis* about the physical nature of light, but a *stipulation* which I can make of my own free will in order to arrive at a definition of simultaneity." This means that the time t'_E is just a different convention. If the time t'_E is the true time, physical time, or nature's choice, why did Einstein say it is a stipulation of his free will? Ironically, after the supremacy of t'_E has been established, it is no longer conventional any more to the relativists. No one else is allowed to make a different stipulation of his free will. Any other time conventions are considered non-physical. In the anti-conventionalist relativist view, the theory of relativity is the absolute truth. In this sense and under this anti-conventionalist interpretation, the theory of relativity is not relative, but it is really absolutism in disguise. Ohanian is one of such anti-conventionalist relativists. He writes [Ohanian (2009), pp. 95, 96, 102]: "he (Einstein) was stipulating something that was not subject to. ... Einstein was entitled to make a hypothesis about the speed of light, but not a stipulation. The speed of light is either constant or not, and only measurement can decide what it is. ... Einstein was lucky... What he had asserted by stipulation actually was confirmed by experiment. In the end, he turned out to have been right for the wrong reason." I disagree with Ohanian on this. In my opinion, debating what is *physical* is like a religious war. If we want to use the word *physical* with a clear meaning, it should mean the part of nature's laws that are independent of human conventions. Time, as well as length measurement standard, is exactly a human convention.

An anti-conventionalist may argue that by "physical" he means the co-ordinates and quantities which can be measured by "physical instruments", and the Newton coordinates and the N-time x'_N, y'_N, z'_N, t'_N in Eq. 10.34 cannot be measured by physical instruments. I do not see why not. If we can define it, we can measure it. The question is not whether we can, but rather whether we want. I have given different ways in the above, of implementing the N-clocks and measuring the N-time. The only difference between x'_N, y'_N, z'_N, t'_N and x'_E, y'_E, z'_E, t'_E is just a matter of convenience.

When Einstein advocated the "principle of general covariance" [Einstein (1916)], he did not classify the general transformations in Eq. 10.31 into two categories of the physical and the non-physical, not to mention giving a criterion to distinguish the physical from the non-physical. Furthermore, Eq. 10.31 indeed includes two different types of transformations. Both the Lorentz transformation and Galilean transformation belong to the first type—transformation between two reference frames. The other type is the following:

$$x' = x'(x, y, z),$$
$$y' = y'(x, y, z),$$
$$z' = z'(x, y, z),$$
$$t' = t'(x, y, z, t). \tag{10.43}$$

(Notice that the space transformations in x', y', z' do not involve time t.) This is only the coordinate and time standard change within the *same reference frame*. Eq. 10.34 belongs to this type. This type of re-coordination may result in a different simultaneity standard for the *same reference frame*. But Eq. 10.43 is just a special case of Eq. 10.31, and Einstein did not single out this type of re-coordination and ban it. If there is a unique time standard (simultaneity standard) for the same reference frame, any non-trivial re-coordination of the type of Eq. 10.43 should be banned and Einstein's "principle of general covariance" would be much more restrictive. If so, it should not even be called "general covariance" any more.

In the context of Schwarzschild spacetime and black holes, various coordinate systems have been studied, including the Schwarzschild coordinates, Kruskal-Szekeres coordinates, tortoise coordinates, Eddington-Finkelstein coordinates, isotropic coordinates, Gullstrand-Painlevé coordinates and Lemaître coordinates, but there has never been any discussion in literature which is physical and which is non-physical.

*** Debate**: Relativity vs. conventionality of simultaneity

At the heart, the debate is "relativity of simultaneity" vs. "conventionality of simultaneity". The two opposing schools agree on that different reference frames may have different simultaneity standards, but the answer to a further question divides the two schools—conventionalist and anti-conventionalist—Can we have different simultaneity standards for the same reference frame K′? My opinion is yes, and I do not see why not. It is not a matter of validity, but rather a matter of convenience.

M. Friedman writes [(1983), p. 165–166]: "What we have just described is the well-known and uncontroversial relativity of simultaneity in Minkowski space-time. Inertial frames or trajectories in motion relative to one another do not agree on simultaneity, so simultaneity has to be relativized to a choice of frame or trajectory. However, a second claim or thesis about simultaneity has insinuated itself into the literature in a very confusing way. This second thesis says nothing about relative motion and the comparison of different inertial frames; rather, it concerns the status of simultaneity within a single inertial frame."

What Friedman fails to realize is that what is controversial or uncontroversial is also relative. What is uncontroversial to one group of people can be controversial to another group, and vice versa. What is controversial or uncontroversial is also relative to time: the idea that the earth was moving was controversial two thousand years ago, as well as four hundred years ago, but it is commonly accepted today. The idea of conventionality of simultaneity (which is the true meaning of relativity of simultaneity) is confusing to Friedman in the same way that Galileo's "heretic" idea that the earth was moving was confusing to the church people four hundred years ago. The interpretation of relativity insisting a unique God-given simultaneity standard per reference frame is really absolutism in disguise. I believe that the conventionality of time will be commonly accepted in the next four hundred years to come, just as heliocentrism is commonly accepted today.

*** Yet Another Paradox**: The heliocentrism-geocentrism paradox (Copernicus-Ptolemy paradox)

Nicolaus Copernicus is credited for his heliocentric revolution against the older geocentric view. However, we can say that the earth moves around the sun, or the sun moves around the earth as well, since motion is relative. This is just a matter what reference frame we choose. Was the Copernican revolution much ado about nothing? I shall call this the heliocentrism-geocentrism paradox, or Copernicus-Ptolemy paradox.

The geocentric model of astronomy was suggested in Ancient Greece by Aristotle. The founder of quantitative mathematical astronomy is Apollonius ($c.$ 200 BCE). The theory was further developed by Hipparchus ($c.$ 130 BCE) and culminated with the *Almagest* of Ptolemy ($c.$ 150 CE). It is commonly known as the Ptolemaic system. In the geocentric model, the earth is at the center of the celestial sphere where all the celestial bodies are located. The celestial sphere rotates around the center in a period of one sidereal day (about 4 minutes shorter than a mean solar day). The stars are fixed on the celestial sphere while the sun moves on a great circle known as the ecliptic with a period of one year. The motions of the planets on the celestial sphere are more complex than the sun, sometimes even in retrograde motion (moving backward). Apollonius introduced the epicycle model to explain the planets motion: a planet P moves at a constant speed v_P on a circle called the epicycle with radius r_e and center S, while S moves with a constant speed v_S on a circle called the deferent with radius r_d and the center at the earth E (see Figure 10.18a). By guessing the relative speeds v_p and v_S, and the relative radii r_e and r_d, these Greek scientists tried to fit the observational astronomical data to explain the motions of the planets.

The result was not extremely satisfactory. Ptolemy then developed more complex models, like epicycle whose center moves on another epicycle, but the model fitting still could not agree completely with observational data. According to Wikipedia:

> In part, due to misunderstandings about how deferent/epicycle models worked, "adding epicycles" has come to be used as a derogatory comment in modern scientific discussion. The term might be used, for example, to describe continuing to try to adjust a theory to make its predictions match the facts. There is a generally accepted idea that extra epicycles were invented to alleviate the growing errors that the Ptolemaic system noted as measurements became more accurate, particularly for Mars. According to this notion, epicycles are regarded by some as the paradigmatic example of bad science.

A remark in [Kolb (1996)] is such an example: "The Copernican system as proposed by Copernicus is neither simpler nor more intuitive, nor does it do a much better job of agreeing with observations. So why, then is Copernicus a hero and 'epicycles' the ultimate pejorative description of an ugly scientific model? It is because Copernicus's model was an enormous step in the right direction. Newton could connect astronomy with the sciences of mechanics and dynamics using Copernicus's model, whereas Ptolemy's model was barren. In this respect, Copernicus had the truer model."

I disagree with Kolb and I have a rather different view. The epicycle model was not successful, not because the geocentric model is bad, but because the Greeks did not find good parameters—the position of the center S, radius of the epicycle r_e, the speed v_P of P and the speed v_S of S—to fit the model to the observational data. In fact, if S is taken as the sun, and the radius r_e as the radius of the planet orbit (of course elliptic epicycle would be more accurate), the epicycle model is the geocentric translation of the heliocentric picture. The radius of the deferent should be the distance between the sun and the earth. Figure 10.18(a) represents the motion of an inferior planet (whose orbit is inside the orbit of the earth). It is also important to note that the epicycle should not be on the celestial sphere, but rather in the ecliptic plane (approximately), which intersects the celestial sphere. For superior planets like Mars, the radius

r_e of the epicycle should be even greater than r_d, the radius of the deferent, as in Figure 10.18(b). This was more difficult for the ancient Greeks to guess. If Apollonius and Ptolemy had found correct parameters, the Ptolemaic system and the Copernican system are equivalent.

By now we see, heliocentrism and geocentrism are just two different but equivalent descriptions of nature. Sometimes the heliocentric reference frame is more convenient, but sometimes the geocentric reference frame is more convenient. Figure 10.19 shows a geocentric view of the diurnal motion of the sun on the celestial sphere. Using this simple geocentric model, we can calculate the night length n and day length d (in hours) at any latitude λ throughout the year [Guo and Mehrubeoglu (2012)]:

$$n = \frac{24}{\pi} \cos^{-1} \left(\tan \lambda \cdot \tan \sigma \right),$$
$$d = 24 - n,$$

where $\sigma = \angle HOQ$ is the solar declination, which is the angle from the sun at noon to the equatorial plane and it can be found from

$$\sin \sigma = \sin \varepsilon \cdot \sin \Phi = \sin \varepsilon \cdot \sin \left(\frac{2\pi D}{Y} \right),$$

where $\varepsilon = 23.5°$ is the obliquity of the ecliptic (angle between the ecliptic plane and the equatorial plane), $\Phi = 2\pi D/Y$ is the ecliptic longitude of the sun, D is the number of days after the vernal equinox, and $Y = 365.24$ is the number of days in a year. The local sunrise time is $t_r = n/2$.

What is the significance of Copernican heliocentric revolution? It demoted the geocentric view from the absolute position as the only possible world view. The heliocentric view and the geocentric view are different but equivalent. One cannot be more true than the other; it can only be more convenient. The difference of the two is only a matter of convenience.

Similarly, Einstein's revolution demoted the Galilean transformation from the absolute position as the only possible time convention. Newton's time convention and Einstein's time convention are different but equivalent. The difference of the two is only a matter of convenience.

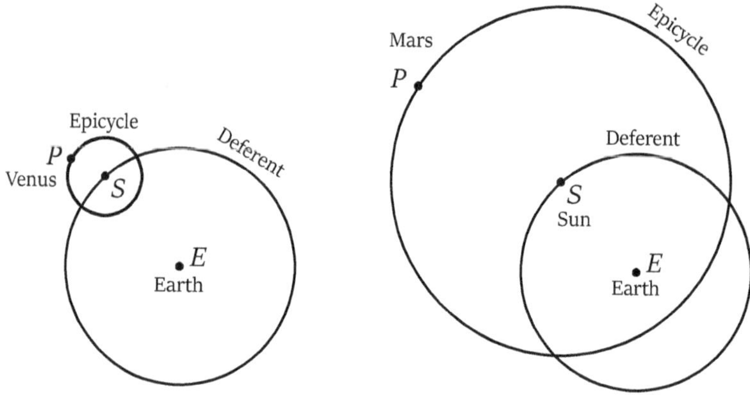

(a) An inferior planet (e.g., Venus)   (b) A superior planet (e.g., Mars)

Figure 10.18   The epicycle model

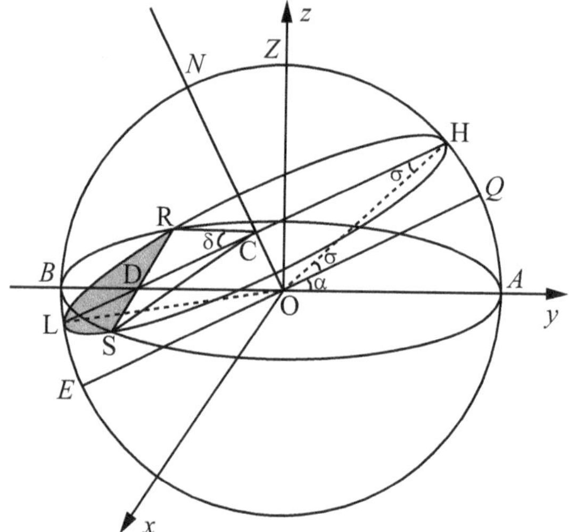

Figure 10.19   The diurnal motion of the sun on the celestial sphere—LRHSL (L: midnight; R: sunrise; H: noon; S: sunset; colatitude $\alpha = \frac{1}{2}\pi - \lambda$) [Guo and Mehrubeoglu (2012)]

*** Philosophical View**: What is time exactly?

—My opinion: Time is the measurement of motion using another motion as standard. Time is a human convention.

If I have invented my law of motion as the following,

$$\mathbf{F} = mt\frac{d}{dt}\left(t\frac{d\mathbf{r}}{dt}\right),$$ (10.44)

for bodies moving in an inertial reference frame, where $\mathbf{F}$ is the force on the body, m the mass, $\mathbf{r}$ the position and t the time, you may say I am wrong. According to my law of motion,

$$x = a\ln t$$ (10.45)

is a solution when $\mathbf{F} = 0$, where a is a constant. This is not uniform motion. The motion slows down with time. This contradicts Newton's first law and second law, which says

$$\mathbf{F} = m\frac{d^2\mathbf{r}}{dt^2}.$$ (10.46)

I would say, my law is the same as Newton's second law, but I only used a time standard different from that of Newton. To distinguish them and avoid confusion, let us use τ for Newton's time. Newton's second law is written as

$$\mathbf{F} = m\frac{d^2\mathbf{r}}{d\tau^2}.$$ (10.47)

My time t is related to τ by

$$t = e^\tau, \text{ or } \tau = \ln t.$$ (10.48)

This time transformation is certainly allowed by Einstein's "principle of general covariance" in general relativity, and it is a special case of Eq. 10.31. If we substitute the time transformation Eq. 10.48 into Eq. 10.47, we obtain my law of motion in Eq. 10.44. I have adopted a clock (t) which runs faster and faster against the Newtonian clock (τ). That is why in my description the motion of a body slows down with time even in an inertial reference frame when there is zero force.

My law of motion is the same as the second law of Newton. They describe the same law of nature and the only difference is a matter of convenience.

From this example we can see, the existence of inertial reference frames, namely the law of inertia, or Newton's first law, is not a wonder in nature which was discovered by Newton. (The concept of inertial frames was due to Ludwig Lange in 1885. Newton's original version is the absolute space, which is even more problematic, because space is not any existence. The meaning of space is the nonexistence of matter. Space without matter cannot be used as a reference system.) It comes to our revelation that the concept of inertial reference frame has a part of human convention of time standard in it.

Proposition 2. *Although the theory of relativity is opposed to Newton's absolutism in time and space, it is actually just another form of absolutism in disguise. If Newton's position is at the north pole, then Einstein's position is at the south pole. If Newton's view is the northern absolutism, then relativity theory is the southern absolutism. Now the southern dynasty of absolutism has overthrown and replaced the northern dynasty.*

Proposition 3.

(1) *Copernicus' revolution demoted geocentrism from the position of the absolute truth. However, his new heliocentrism should not replace geocentrism to become the new absolute truth.*

(2) *Lobachevsky's revolution demoted Euclidean geometry from the position of the absolute truth. However, his non-Euclidean geometry should not replace Euclidean geometry to become the new absolute truth.*

(3) *Einstein's revolution demoted Newton's time convention from the position of the absolute truth. However, his new time convention should not replace Newton's time convention to become the new absolute truth.*

Proposition 4. *One time standard cannot be more true than another; it can only be more convenient.*

Proposition 5. *Time is the measurement of motion using another motion as standard. Time is a human convention. There is no true time, or God-given time, or by whatever other names—natural time, physical time, cosmic time, etc. The phrase "physical time" is an oxymoron.*

Proposition 6. *The essence of the theory of special relativity is to use the motion of light as the time standard for every inertial reference system. The constancy of the speed of light is the tautological consequence of this time convention, rather than a new law of nature discovered by Einstein.*

Proposition 7. *God created matter.*[10] *Man created time. Gravity does not curve spacetime. Man curves it.*

---

[10]The word "God" throughout this book is used as a figure of speech. It means "nature", and has nothing to do with religious beliefs.

Appendix 1. Topics of Linear Algebra

§1. Proof of Commutativity of Addition

For the reader's convenience, we copy Definition 11 in Chap. 1 to here.

Definition 1. Let F be a field and V a nonempty set. V together with two operations called addition and scalar-vector multiplication is called a vector space, if for all $\mathbf{u}, \mathbf{v}, \mathbf{w} \in V$ and $a, b \in F$, the following conditions are satisfied.

(1) $(\mathbf{u} + \mathbf{v}) + \mathbf{w} = \mathbf{u} + (\mathbf{v} + \mathbf{w})$;

(2) There exists $\mathbf{0} \in V$ such that $\mathbf{v} + \mathbf{0} = \mathbf{v}$;

(3) For any $\mathbf{v} \in V$, there exists $\mathbf{x} \in V$ such that $\mathbf{v} + \mathbf{x} = \mathbf{0}$. We denote $\mathbf{x} = -\mathbf{v}$;

(4) $a(\mathbf{u} + \mathbf{v}) = a\mathbf{u} + a\mathbf{v}$;

(5) $(a + b)\mathbf{v} = a\mathbf{v} + b\mathbf{v}$;

(6) $a(b\mathbf{v}) = (ab)\mathbf{v}$;

(7) $1\mathbf{v} = \mathbf{v}$, where $1 \in F$ is the multiplicative identity in F.

Lemma 1. $-\mathbf{v} + \mathbf{v} = \mathbf{0}$.
Proof .

$$
\begin{aligned}
-\mathbf{v} + \mathbf{v} &= -\mathbf{v} + \mathbf{v} + \mathbf{0} && \text{(axiom 2)}\\
&= -\mathbf{v} + \mathbf{v} + \{(-\mathbf{v}) + [-(-\mathbf{v})]\} && \text{(axiom 3)}\\
&= -\mathbf{v} + [\mathbf{v} + (-\mathbf{v})] + [-(-\mathbf{v})] && \text{(axiom 1)}\\
&= (-\mathbf{v} + \mathbf{0}) + [-(-\mathbf{v})] && \text{(axiom 3)}\\
&= -\mathbf{v} + [-(-\mathbf{v})] && \text{(axiom 2)}\\
&= \mathbf{0}. && \text{(axiom 3)}
\end{aligned}
$$

Lemma 2. $\mathbf{0} + \mathbf{v} = \mathbf{v}$.
Proof .

$$
\begin{aligned}
\mathbf{0} + \mathbf{v} &= [\mathbf{v} + (-\mathbf{v})] + \mathbf{v} && \text{(axiom 3)}\\
&= \mathbf{v} + [(-\mathbf{v}) + \mathbf{v}] && \text{(axiom 1)}\\
&= \mathbf{v} + \mathbf{0} && \text{(lemma 1)}\\
&= \mathbf{v}. && \text{(axiom 2)}
\end{aligned}
$$

Theorem 1. [(Bryant (1971)] $\mathbf{u} + \mathbf{v} = \mathbf{v} + \mathbf{u}$.

Proof.

$$
\begin{aligned}
\mathbf{u} + \mathbf{v} &= \mathbf{0} + \mathbf{u} + \mathbf{v} + \mathbf{0} && \text{(lemma 1, axiom 2)}\\
&= (-\mathbf{u} + \mathbf{u}) + \mathbf{u} + \mathbf{v} + [\mathbf{v} + (-\mathbf{v})] && \text{(lemma 1, axiom 3)}\\
&= -\mathbf{u} + (\mathbf{u} + \mathbf{u} + \mathbf{v} + \mathbf{v}) + (-\mathbf{v}) && \text{(axiom 1)}\\
&= -\mathbf{u} + (1\mathbf{u} + 1\mathbf{u} + 1\mathbf{v} + 1\mathbf{v}) + (-\mathbf{v}) && \text{(axiom 7)}\\
&= -\mathbf{u} + 2\mathbf{u} + 2\mathbf{v} + (-\mathbf{v}) && \text{(axiom 5)}\\
&= -\mathbf{u} + 2(\mathbf{u} + \mathbf{v}) + (-\mathbf{v}) && \text{(axiom 4)}\\
&= -\mathbf{u} + (\mathbf{u} + \mathbf{v}) + (\mathbf{u} + \mathbf{v}) + (-\mathbf{v}) && \text{(axiom 5, 7)}\\
&= (-\mathbf{u} + \mathbf{u}) + \mathbf{v} + \mathbf{u} + [\mathbf{v} + (-\mathbf{v})] && \text{(axiom 1)}\\
&= \mathbf{v} + \mathbf{u}. && \text{(lemma 1,2, axiom 3, 2)}
\end{aligned}
$$

Definition 1 can also be put in an equivalent form with six independent axioms [Rigby and Wiegold (1973)] as follows. Note there is no need for the existential quantifier $\exists$ or a constant symbol $\mathbf{0}$ for the zero vector in these axioms.

Definition 2. Let F be a field and V a nonempty set. V together with addition and scalar-vector multiplication is called a vector space if for all $\mathbf{u}, \mathbf{v} \in V$ and $a, b \in F$, the following conditions are satisfied.

(1) $(\mathbf{u} + \mathbf{v}) + \mathbf{w} = \mathbf{u} + (\mathbf{v} + \mathbf{w})$;
(2) $a(\mathbf{u} + \mathbf{v}) = a\mathbf{u} + a\mathbf{v}$;

(3) $(a+b)\mathbf{v} = a\mathbf{v} + b\mathbf{v}$;

(4) $a(b\mathbf{v}) = (ab)\mathbf{v}$;

(5) $0\mathbf{u} = 0\mathbf{v}$;

(6) $1\mathbf{v} = \mathbf{v}$.

§2. Covectors and the Dual Space

Definition 3. A linear mapping $f : V \to F$ is called a **linear function** (or **linear functional**, or **linear form**). Let f_1, f_2 be linear functions and $a \in F$. We define the addition of f_1 and f_2 to be a linear function f such that

$$f(\mathbf{v}) = f_1(\mathbf{v}) + f_2(\mathbf{v}),$$

and we define the multiplication of a and f_1 to be g such that

$$g(\mathbf{v}) = af_1(\mathbf{v}).$$

Definition 4. Let V be a vector space over a field F. All the linear functions (or linear forms) form a vector space and it is called the **dual space** of V, denoted by V^*.

The vectors in V^* are also called **covectors** (or **covariant vectors**). By contrast, the vectors in V are called **contravariant vectors**. We will see the reason for this naming in the next section when an inner product is introduced and an isomorphism between V and V^* is established with the help of the inner product.

Let $(x, y, z) \in V = \mathbb{R}^3$. A linear function $f \in V^*$ maps (x, y, z) to a real number. In general, $f(x, y, z) = a_1 x + a_2 y + a_3 z$, where $a_1, a_2, a_3 \in \mathbb{R}$. Here the linear function coincide with the linear function in the sense of polynomial function of degree one. We can see here the linear function f can be uniquely represented by a 3-tuple (a_1, a_2, a_3). So f itself is a vector of dimension 3.

Definition 5. Let $\{\mathbf{e}_1, \ldots, \mathbf{e}_n\}$ be a basis of vector space V. We define a set of linear functions $f_1, \ldots, f_n$ as follows: for any vector $\mathbf{v} = a_1\mathbf{e}_1 + \cdots + a_n\mathbf{e}_n \in V$, we define

$$f_i \mathbf{v} \stackrel{\text{def}}{=} a_i, \quad i = 1, \ldots, n.$$

Basically f_i is the projection operator that maps a vector $\mathbf{v}$ to its ith coordinate under a given basis. This set of linear functions $\{f_1, \ldots, f_n\}$ is called the (affine) **dual basis** of $\{\mathbf{e}_1, \ldots, \mathbf{e}_n\}$.

The dual basis so defined is called the **affine dual basis**, to distinguish from the other dual basis, the **metric dual basis** established through an inner product.

It needs to be justified that the "dual basis" defined above is indeed a basis for V^*. Let f be any linear function. Suppose the images of f on the basis vectors $\{\mathbf{e}_1, \ldots, \mathbf{e}_n\}$ are

$$f\mathbf{e}_i = \tau_i, \ i = 1, \ldots, n.$$

For any vector $\mathbf{v} = a_1\mathbf{e}_1 + \ldots + a_n\mathbf{e}_n$, we have

$$\begin{aligned}
f\mathbf{v} &= f(a_1\mathbf{e}_1 + \cdots + a_n\mathbf{e}_n) \\
&= a_1 f(\mathbf{e}_1) + \cdots + a_n f(\mathbf{e}_n) \\
&= a_1\tau_1 + \cdots + a_n\tau_n \\
&= \tau_1 f_1(\mathbf{v}) + \cdots + \tau_n f_n(\mathbf{v}) \\
&= (\tau_1 f_1 + \cdots + \tau_n f_n)(\mathbf{v}).
\end{aligned}$$

This means that f is a linear combination of f_i, $i = 1, 2, \ldots, n$. Namely,

$$f = \tau_1 f_1 + \cdots + \tau_n f_n.$$

The dual space V^* also has dimension n. Since all the vector spaces of the same dimension are isomorphic to each other, V^* is isomorphic to V.

If two vector spaces are isomorphic to each other, there are infinitely many different isomorphisms between them. After choosing a basis $\{\mathbf{e}_1, \ldots, \mathbf{e}_n\}$ for V, an isomorphism $\Psi : V \to V^*$ can be easily constructed. Just define the isomorphism on the basis, $\Psi(\mathbf{e}_i) = f_i$, $i = 1, \ldots, n$, and the mapping can be linearly extended to the entire space V. Since the dual basis is constructed based on projection, we call this isomorphism Ψ **affine duality mapping**. A vector $\mathbf{v} \in V$ and $\Psi(\mathbf{v}) \in V^*$ are called **affine dual** of each other. $\Psi(\mathbf{v})$ is also denoted by $\mathbf{v}^*$ and thus $\mathbf{v}^{**} = \mathbf{v}$. If $\mathbf{v} \in V$ has coordinates $(x_1, \ldots, x_n)$ under basis $\{\mathbf{e}_i\}$, then $f = \Psi(\mathbf{v})$ has the same coordinates $(x_1, \ldots, x_n)$ under dual basis $\{f_i\}$. Therefore under affine duality mapping, vectors and covectors with the same coordinates under respective dual bases are identified as the same.

Suppose $\mathbf{v}$ is any vector in V and f is any linear function such that $f(\mathbf{v}) = a \in F$. $\mathbf{v}$ can also be viewed as a linear function on V^*. Namely, $\mathbf{v} : V^* \to F$ such that for any $f \in V^*$, we define $\mathbf{v}(f) \overset{\text{def}}{=} f(\mathbf{v}) = a$.

§3. Inner Product

Definition 6. Let V be a real vector space. A mapping $\langle \cdot, \cdot \rangle : V \times V \to \mathbb{R}$; $(\mathbf{u}, \mathbf{v}) \mapsto \langle \mathbf{u}, \mathbf{v} \rangle$ is called a (real) **inner product** (or **dot product**), if it satisfies the following conditions, for all $\mathbf{u}, \mathbf{v}, \mathbf{u}_1, \mathbf{u}_2, \mathbf{v}_1, \mathbf{v}_2 \in V$ and $a_1, a_2 \in \mathbb{R}$.

(1) Bilinear:

   (1a). $\langle a_1 \mathbf{u}_1 + a_2 \mathbf{u}_2, \mathbf{v} \rangle = a_1 \langle \mathbf{u}_1, \mathbf{v} \rangle + a_2 \langle \mathbf{u}_2, \mathbf{v} \rangle$.

   (1b). $\langle \mathbf{u}, a_1 \mathbf{v}_1 + a_2 \mathbf{v}_2 \rangle = a_1 \langle \mathbf{u}, \mathbf{v}_1 \rangle + a_2 \langle \mathbf{u}, \mathbf{v}_2 \rangle$.

(2) Symmetric: $\langle \mathbf{v}_1, \mathbf{v}_2 \rangle = \langle \mathbf{v}_2, \mathbf{v}_1 \rangle$.

(3) Positive-definite: $\langle \mathbf{v}, \mathbf{v} \rangle \geq 0$. Furthermore, $\langle \mathbf{v}, \mathbf{v} \rangle = 0 \to \mathbf{v} = 0$.

Alternatively, $\langle \mathbf{u}, \mathbf{v} \rangle$ is also denoted by $\mathbf{u} \cdot \mathbf{v}$. Note, (1b) is not independent and can be derived from (1a) and (2).

A real vector space V together with an inner product $\langle \cdot, \cdot \rangle : V \times V \to \mathbb{R}$ is called a real **inner product space**. A finite dimensional real inner product space is called a **Euclidean Space**.

For any vectors $\mathbf{u}, \mathbf{v} \in V$, let $\{\mathbf{e}_1, \ldots, \mathbf{e}_n\}$ be a basis for V and $\mathbf{u} = x_1 \mathbf{e}_1 + \cdots + x_n \mathbf{e}_n$ and $\mathbf{v} = y_1 \mathbf{e}_1 + \cdots + y_n \mathbf{e}_n$. Suppose $\langle \cdot, \cdot \rangle$ is any inner product. We have

$$\langle \mathbf{u}, \mathbf{v} \rangle = \left\langle \sum_{i=1}^{n} x_i \mathbf{e}_i, \sum_{j=1}^{n} y_j \mathbf{e}_j \right\rangle$$

$$= \sum_{i=1}^{n} \sum_{j=1}^{n} x_i y_j \langle \mathbf{e}_i, \mathbf{e}_j \rangle$$

$$= \sum_{i=1}^{n} \sum_{j=1}^{n} x_i y_j g_{ij}$$

where

$$g_{ij} = \langle \mathbf{e}_i, \mathbf{e}_j \rangle$$

are real numbers which form a matrix, denoted by $[g]$. This can be written in the matrix form

$$\langle \mathbf{u}, \mathbf{v} \rangle = \begin{bmatrix} x_1 & \cdots & x_n \end{bmatrix} [g] \begin{bmatrix} y_1 \\ \vdots \\ y_n \end{bmatrix}$$

$$= \begin{bmatrix} x_1 & \cdots & x_n \end{bmatrix} \begin{bmatrix} g_{11} & \cdots & g_{1n} \\ \vdots & \vdots & \vdots \\ g_{n1} & \cdots & g_{nn} \end{bmatrix} \begin{bmatrix} y_1 \\ \vdots \\ y_n \end{bmatrix} .$$

The matrix $[g]$ is called the **metric matrix** for inner product $\langle \cdot, \cdot \rangle$. $[g]$ must be a symmetric matrix because $\langle \mathbf{e}_i, \mathbf{e}_j \rangle = \langle \mathbf{e}_j, \mathbf{e}_i \rangle$. Each inner product has a **matrix representation**. An inner product is uniquely determined by its metric matrix.

Two vectors $\mathbf{u}$ and $\mathbf{v}$ are said to be **orthogonal** to each other, if $\langle \mathbf{u}, \mathbf{v} \rangle = 0$. The concept of orthogonal is the generalization of the geometrical concept of perpendicular. In general, we can define the lengths of vectors and angles between two vectors.

For any $\mathbf{v} \in V$, we define the **length** (or the **norm**) of $\mathbf{v}$ to be

$$\|\mathbf{v}\| \stackrel{\text{def}}{=} \sqrt{\langle \mathbf{v}, \mathbf{v} \rangle}.$$

For any $\mathbf{u}, \mathbf{v} \in V$, the distance between $\mathbf{u}$ and $\mathbf{v}$ induced by the inner product is defined to be

$$d(\mathbf{u}, \mathbf{v}) \stackrel{\text{def}}{=} \|\mathbf{u} - \mathbf{v}\| = \sqrt{\langle \mathbf{u} - \mathbf{v}, \mathbf{u} - \mathbf{v} \rangle}.$$

The angle between $\mathbf{u}$ and $\mathbf{v}$ is defined to be

$$\theta \stackrel{\text{def}}{=} \cos^{-1} \frac{\langle \mathbf{u}, \mathbf{v} \rangle}{\|\mathbf{u}\| \, \|\mathbf{v}\|}.$$

§4. Contravariant and Covariant Components of Vectors

Definition 7. Let $\hat{\mathbf{e}}_1, \ldots, \hat{\mathbf{e}}_n \in V$ be defined as

$$\langle \hat{\mathbf{e}}_i, \mathbf{e}_j \rangle \stackrel{\text{def}}{=} f_i \mathbf{e}_j = \delta_{ij}.$$

Then $\{\hat{\mathbf{e}}_1, \ldots, \hat{\mathbf{e}}_n\}$ is a basis for V, and we call it **reciprocal basis** of $\{\mathbf{e}_1, \ldots, \mathbf{e}_n\}$.

Theorem 2. *The reciprocal basis and the original basis are related by*

$$\hat{\mathbf{e}}_i = \sum_{k=1}^{n} g^{ik} \mathbf{e}_k,$$

$$\mathbf{e}_i = \sum_{k=1}^{n} g_{ik} \hat{\mathbf{e}}_k.$$

If the dimension of the vector space is 3, then the reciprocal basis of $\{\mathbf{e}_1, \mathbf{e}_2, \mathbf{e}_3\}$ *in* $\mathbb{R}^3$ *can be expressed as*

$$\hat{\mathbf{e}}_1 = \frac{\mathbf{e}_2 \times \mathbf{e}_3}{\mathbf{e}_1 \cdot (\mathbf{e}_2 \times \mathbf{e}_3)}$$

$$\hat{\mathbf{e}}_2 = \frac{\mathbf{e}_3 \times \mathbf{e}_1}{\mathbf{e}_1 \cdot (\mathbf{e}_2 \times \mathbf{e}_3)}$$

$$\hat{\mathbf{e}}_3 = \frac{\mathbf{e}_1 \times \mathbf{e}_2}{\mathbf{e}_1 \cdot (\mathbf{e}_2 \times \mathbf{e}_3)}.$$

It is easy to verity that $\langle \hat{\mathbf{e}}_i, \mathbf{e}_j \rangle = \delta_{ij}$ for $i, j = 1, 2, 3$.

The vectors in a vector space V are also called contravariant vectors. The vectors in the dual space V^*, namely the linear functions on V, are also called covariant vectors. The reason for these names is that in a bases change, the transformation law for contravariant vectors involve the inverse matrix A^{-1} of the transition matrix A, while the transformation law for covariant vectors involve the transition matrix A itself.

The reciprocal basis provides a different perspective to view the covariant vectors. A vector $\mathbf{v} \in V$ can be represented by its components $(x^1, \ldots, x^n)$ under the original basis $\{\mathbf{e}_1, \ldots, \mathbf{e}_n\}$,

$$\mathbf{v} = x^1 \mathbf{e}_1 + \cdots + x^n \mathbf{e}_n = \sum_{k=1}^{n} x^k \mathbf{e}_k.$$

The same vector $\mathbf{v}$ can also be represented by its components $(x_1, \ldots, x_n)$ under the reciprocal basis $\{\hat{\mathbf{e}}_1, \ldots, \hat{\mathbf{e}}_n\}$,

$$\mathbf{v} = x_1 \hat{\mathbf{e}}_1 + \cdots + x_n \hat{\mathbf{e}}_n = \sum_{k=1}^{n} x_k \hat{\mathbf{e}}_k.$$

Definition 8. The components $(x^1, \ldots, x^n)$ of vector $\mathbf{v} \in V$ under the original basis $\{\mathbf{e}_1, \ldots, \mathbf{e}_n\}$ are called **contravariant components** of $\mathbf{v}$. The components $(x_1, \ldots, x_n)$ of $\mathbf{v}$ under the reciprocal basis $\{\hat{\mathbf{e}}_1, \ldots, \hat{\mathbf{e}}_n\}$ are called the **covariant components** of $\mathbf{v}$.

Theorem 3. *The covariant components* $(x_1, \ldots, x_n)$ *and contravariant components* $(x^1, \ldots, x^n)$ *of the same vector are related by*

$$x_i = \sum_{k=1}^{n} g_{ik} x^k,$$

$$x^i = \sum_{k=1}^{n} g^{ik} x_k.$$

In the case of orthonormal basis, the reciprocal basis coincides with the original basis and therefore the covariant components also coincide with the contravariant components of the same vector.

Suppose we have a basis change from $\{\mathbf{e}_1, \ldots, \mathbf{e}_n\}$ to $\{\bar{\mathbf{e}}_1, \ldots, \bar{\mathbf{e}}_n\}$ and

$$\bar{\mathbf{e}}_i = \sum_{k=1}^{n} A_i^k \mathbf{e}_k,$$

where A_i^k is the element at the kth row and ith column in the transition matrix A. The new contravariant components under basis $\{\bar{\mathbf{e}}_1, \ldots, \bar{\mathbf{e}}_n\}$ are

$$\bar{x}^i = \sum_{k=1}^{n} \left[A^{-1}\right]^i_k x^k.$$

The new basis $\{\bar{\mathbf{e}}_1, \ldots, \bar{\mathbf{e}}_n\}$ induces a new reciprocal basis $\{\hat{\bar{\mathbf{e}}}_1, \ldots, \hat{\bar{\mathbf{e}}}_n\}$. The new covariant components under basis $\{\hat{\bar{\mathbf{e}}}_1, \ldots, \hat{\bar{\mathbf{e}}}_n\}$ are

$$\bar{x}_i = \sum_{k=1}^{n} A_i^k x_k.$$

Remark. According to K. Reich [(1994)], J. Sylvester introduced the terms "covariant" and "contravariant" in 1851 [Sylvester (1851)]. The naming of "contravariant" and "covariant" is with respect to the transition matrix A of basis transformation. The transformation of the covariant components $\bar{x}_i$ involves the same matrix A, while the transformation of the contravariant components $\bar{x}^i$ involves (the transpose of) the inverse of matrix A. If we call the transformation of basis with matrix A the "forward" transformation, then the transformation of the contravariant components $\bar{x}^i$ are the "backward" transformation, with an analogy: if one rides on the train and the train moves forward, the trees outside seem to move backward.

Remark. The contravariant coordinates (or contravariant components) of a vector are the parallel projections. The covariant coordinates (or covariant components) of a vector are the perpendicular (orthogonal) projections.

We set up a coordinate system Oxy in the plane (Figure 0.1). In general, the two axes Ox and Oy are not orthogonal to each other. They make an angle α.

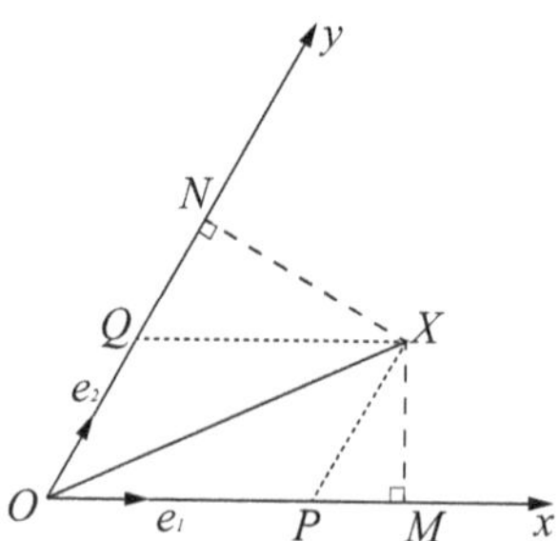

Figure 0.1   Contravariant and covariant components of a vector

Suppose $\mathbf{e}_1$ has unit length on Ox and $\mathbf{e}_2$ has unit length on Oy. X is a point in the plane. There are two different ways we can equip point X with coordinates—contravariant coordinates or covariant coordinates.

4.1 *Contravariant coordinates as the parallel projections*

Theorem 4. *Make $XP \parallel Oy$ and $XQ \parallel Ox$. Then the parallel projections (x^1, x^2) are the contravariant coordinates (also called affine coordinates because we only used parallelism but not orthogonality), where*

$$x^1 = OP,$$
$$x^2 = OQ,$$

with OP and OQ denoting the signed lengths of segments OP and OQ respectively.

4.2 *Covariant coordinates as the perpendicular projections*

Suppose an inner product is also defined in vector space V and let $\langle \mathbf{v}_1, \mathbf{v}_2 \rangle$ denote the inner product of two vectors. The matrix of the inner product is

$$[g] \stackrel{\text{def}}{=} \begin{bmatrix} g_{11} \; g_{12} \\ g_{21} \; g_{22} \end{bmatrix} \stackrel{\text{def}}{=} \begin{bmatrix} \langle \mathbf{e}_1, \mathbf{e}_1 \rangle \; \langle \mathbf{e}_1, \mathbf{e}_2 \rangle \\ \langle \mathbf{e}_2, \mathbf{e}_1 \rangle \; \langle \mathbf{e}_2, \mathbf{e}_2 \rangle \end{bmatrix} = \begin{bmatrix} 1 & \cos\alpha \\ \cos\alpha & 1 \end{bmatrix}.$$

With the inner product, we have the concept of orthogonality.

Theorem 5. *We draw $XM \perp Ox$ and $XN \perp Oy$, then the perpendicular projections (OM, ON) are the covariant coordinates for point X (Figure 0.1):*

$$x_1 = OM = \langle \mathbf{v}, \mathbf{e}_1 \rangle,$$
$$x_2 = ON = \langle \mathbf{v}, \mathbf{e}_2 \rangle.$$

Furthermore, we find

$$x_1 = \langle \mathbf{v}, \mathbf{e}_1 \rangle = \begin{bmatrix} x^1, x^2 \end{bmatrix} \begin{bmatrix} g_{11} \; g_{12} \\ g_{21} \; g_{22} \end{bmatrix} \begin{bmatrix} 1 \\ 0 \end{bmatrix} = g_{11}x^1 + g_{21}x^2.$$

$$x_2 = \langle \mathbf{v}, \mathbf{e}_2 \rangle = \begin{bmatrix} x^1, x^2 \end{bmatrix} \begin{bmatrix} g_{11} \; g_{12} \\ g_{21} \; g_{22} \end{bmatrix} \begin{bmatrix} 0 \\ 1 \end{bmatrix} = g_{21}x^1 + g_{22}x^2.$$

We recognize OM and ON are exactly the covariant coordinates (x_1, x_2).

Another way to see this is to find out the reciprocal basis $\{\hat{\mathbf{e}}_1, \hat{\mathbf{e}}_2\}$ explicitly. Draw $Ox' \perp \mathbf{e}_2$ and mark $\hat{\mathbf{e}}_1$ on the line of Ox'. Similarly, draw

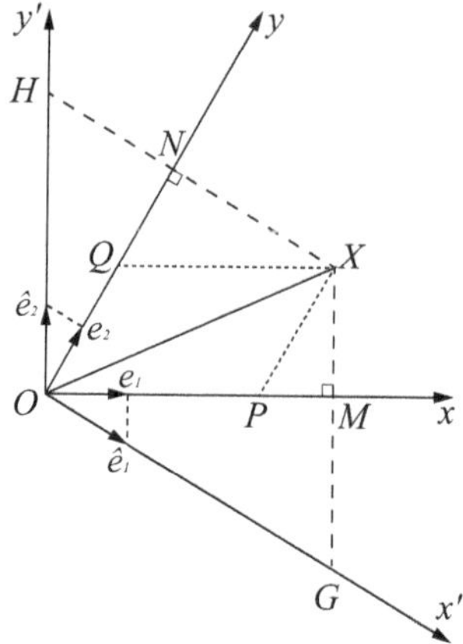

Figure 0.2   Covariant components and reciprocal basis

$Oy' \perp Ox$ and mark $\hat{\mathbf{e}}_2$ on the line of Oy', as illustrated in Figure 0.2. Because $\langle \hat{\mathbf{e}}_1, \mathbf{e}_1 \rangle = 1$ and we have assumed $\|\mathbf{e}_1\| = 1$, we see that $\mathbf{e}_1$ is the orthogonal projection of $\hat{\mathbf{e}}_1$ onto the x-axis. Therefore,

$$\frac{OG}{\|\hat{\mathbf{e}}_1\|} = \frac{OM}{\|\mathbf{e}_1\|} = OM = x_1.$$

Similarly,

$$\frac{OH}{\|\hat{\mathbf{e}}_2\|} = \frac{OM}{\|\mathbf{e}_2\|} = ON = x_2.$$

This means, from X, if we draw $XG \parallel \hat{\mathbf{e}}_2$, and measure OG with $\hat{\mathbf{e}}_1$, we obtain coordinate $x_1 = OM$. If we draw $XH \parallel \hat{\mathbf{e}}_1$, and measure OH with $\hat{\mathbf{e}}_2$, we obtain coordinate $x_2 = ON$. So (x_1, x_2) are the orthogonal projections onto $\mathbf{e}_1$ and $\mathbf{e}_2$ respectively.

§5. Bilinear Forms and Quadratic Forms

Bilinear forms and quadratic forms are important in defining additional structures in vector spaces. If the bilinear form is symmetric and positive-definite, the space is called an **inner product space** (a finite dimensional inner product space is called a **Euclidean space**). When the positive-definite condition is weakened to nondegenerate, it is called a **Minkowski space**. Riemannian manifold is a differentiable manifold whose tangent space at each point is an inner product space. Pseudo-Riemannian manifold is a differentiable manifold whose tangent space at each point is a Minkowski space.

A bilinear form is defined to be a bilinear mapping $\Phi : V \times V \to F$. Let $\{\mathbf{e}_1, \ldots, \mathbf{e}_n\}$ be a basis for V. For $\mathbf{x} = x_1\mathbf{e}_1 + \cdots + x_n\mathbf{e}_n$ and $\mathbf{y} = y_1\mathbf{e}_1 + \cdots + y_n\mathbf{e}_n$,

$$\Phi(\mathbf{x}, \mathbf{y}) = \sum_{i,j=1}^{n} a_{ij} x_i y_j.$$

The matrix $[a_{ij}]$ is called the **matrix associated with the bilinear form** with respect to basis $\{\mathbf{e}_1, \ldots, \mathbf{e}_n\}$.

Definition 9. A bilinear form $\Phi : V \times V \to F$ is said to be **degenerate**, if there exists $\mathbf{v} \neq 0 \in V$, such that for all $\mathbf{x} \in V$, $\Phi(\mathbf{v}, \mathbf{x}) = 0$, or for all $\mathbf{x} \in V$, $\Phi(\mathbf{x}, \mathbf{v}) = 0$. If B is not degenerate, then Φ is said to be **nondegenerate**.

Equivalently, Φ is called nondegenerate if for all $\mathbf{x} \in V$ $\Phi(\mathbf{v}, \mathbf{x}) = 0 \to \mathbf{v} = 0$, and for all $\mathbf{x} \in V$ $\Phi(\mathbf{x}, \mathbf{v}) = 0 \to \mathbf{v} = 0$.

Definition 10. Given a vector space V over a field F, a quadratic form is a mapping $Q : V \to F$ satisfying the following two conditions.
(1) $Q(a\mathbf{x}) = a^2 Q(\mathbf{x})$ for all $a \in F$ and $\mathbf{x} \in V$.
(2) $\Phi(\mathbf{x}, \mathbf{y}) \overset{\text{def}}{=} Q(\mathbf{x} + \mathbf{y}) - Q(\mathbf{x}) - Q(\mathbf{y})$ is a symmetric bilinear form on V.
   Φ is called the symmetric bilinear form associated with Q.

This definition is equivalent to the old-fashioned definition of a quadratic form, as a quadratic homogeneous polynomial with coefficients in a field F:

$$Q(x_1, \ldots, x_n) = \sum_{i,j=1}^{n} a_{ij} x_i x_j,$$

where the matrix $[a_{ij}]$ is a symmetric matrix. The symmetric matrix $[a_{ij}]$ is called the **matrix associated with quadratic form** Q.

Conversely, given any symmetric bilinear form Φ, $Q(\mathbf{x}) \overset{\text{def}}{=} \Phi(\mathbf{x}, \mathbf{x})$ is a quadratic form, called the quadratic form associated with Φ.

Definition 11. A quadratic form Q is said to be **positive-definite**, if for all $\mathbf{x} \neq 0$, $Q(\mathbf{x}) > 0$. Q is said to be **negative-definite**, if for all $\mathbf{x} \neq 0$, $Q(\mathbf{x}) < 0$. Q is said to be **indefinite**, if it is neither positive-definite nor negative-definite. A quadratic form Q is said to be **degenerate** (**nondegenerate**), if the associated symmetric bilinear form Φ is degenerate (nondegenerate).

§6. Free Vector Spaces and Free Algebras

6.1 Intuitive Idea

The "free vector space generated by a set X" is a clever way to make the idea of "formal linear combinations" rigorous. First, let us look at an example. Let $X = \{\mathbf{a}, \mathbf{b}, \mathbf{c}\}$ with three letters $\mathbf{a}, \mathbf{b}, \mathbf{c}$. We may write the "formal linear combinations" of $\mathbf{a}, \mathbf{b}, \mathbf{c}$. Each of these formal linear combinations is called a vector. For example,

$$\mathbf{v}_1 = 2\mathbf{a} + \mathbf{b} + 3\mathbf{c},$$
$$\mathbf{v}_2 = \mathbf{a} - 2\mathbf{b} + \mathbf{c}.$$

We define the addition of two vectors $\mathbf{v}_1$ and $\mathbf{v}_2$ as a formal linear combination of $\mathbf{a}, \mathbf{b}, \mathbf{c}$ by combining like terms,

$$\mathbf{v}_1 + \mathbf{v}_2 = 3\mathbf{a} - \mathbf{b} + 4\mathbf{c}.$$

For a vector $\mathbf{v}_1$ and a scalar, for example 3, we define the scalar multiplication by using the distributive law,

$$3\mathbf{v}_1 = 6\mathbf{a} + 3\mathbf{b} + 9\mathbf{c}.$$

Then all these formal finite linear combinations form a vector space, called the free vector space generated by a set X and is denoted by $\mathscr{V}_F \langle X \rangle$, where F is the field in which all the coefficients are drawn from. The free vector space $\mathscr{V}_{\mathbb{R}} \langle X \rangle$ is a three dimensional vector space, which is isomorphic to $\mathbb{R}^3$ if $F = \mathbb{R}$. The set $X = \{\mathbf{a}, \mathbf{b}, \mathbf{c}\}$ naturally becomes a basis for $\mathscr{V}_{\mathbb{R}} \langle X \rangle$.

The following is the formal definition that captures this idea.

6.2 Formal Definition of Free Vector Space

Definition 12. Let X be a nonempty set and F be a field. The set $\mathscr{V}_F \langle X \rangle$ of vectors is defined to be all the functions $f : X \to F$ that have non-zero values only on finitely many points of X. For $f, g \in \mathscr{V}_F \langle X \rangle$ and $a \in F$, we define the addition $f + g$ and scalar multiplication as follows: for all $x \in X$, define $(f + g)(x) \overset{\text{def}}{=} f(x) + g(x)$, and $(af)(x) \overset{\text{def}}{=} af(x)$. $\mathscr{V}_F \langle X \rangle$ forms a vector space over F. It is called the **free vector space generated by set** X.

A function $f : X \to F$ is said to have **finite support**, if it takes non-zero values only on finitely many points of X.

For each $x \in X$, we define a function $f_x : X \to F$ such that for any $y \in X$,

$$f_x(y) = \begin{cases} 1 & \text{if } y = x \\ \\ 0 & \text{if } y \neq x. \end{cases}$$

For any function $f \in \mathscr{V}_F \langle X \rangle$, suppose f takes non-zero values at point $x_1, ..., x_n$. Namely $f(x_1) = a_1, ..., f(x_n) = a_n$. Then f can be written as

$$f = a_1 f_{x_1} + ... + a_n f_{x_n}.$$

Hence the set of functions $\{f_x \mid x \in X\}$ forms a basis for $\mathscr{V}_F \langle X \rangle$. There is a one-to-one correspondence between the elements $x \in X$ and the basis vectors $f_x \in \mathscr{V}_F \langle X \rangle$. When confusion is not feared, we identify x with f_x and hence the set X is exactly a basis for $\mathscr{V}_F \langle X \rangle$. The dimension of $\mathscr{V}_F \langle X \rangle$ is the cardinality of X. If X is a finite set, then $\mathscr{V}_F \langle X \rangle$ is a finite dimensional vector space. Otherwise, it is infinite dimensional.

Remark. A free vector space generated by a set X is a special case of a free module generated by a set X. Modules are generalizations of vector spaces. Not all the modules have a basis while every vector space has a basis. A module that has a basis is called a free module. All the vector spaces are free in this sense. A "free-module-generated-by-a-set-X" is a free module in this sense and the set X is a natural basis for the module.

6.3 Free Algebras

Given any nonempty set X, we can construct an algebra, called the **free associative algebra generated by** X, based on the construction of the free vector space generated by X. To construct an algebra, we need to define multiplication of two vectors. We use the juxtaposition (or concatenation) idea from Gibbs.

Let X be any nonempty set. The elements in X are considered letters and X is called an alphabet. X can be a finite or infinite set. Let X^* be the set of all strings (finite sequences) over the alphabet X. Precisely,

$$X^* = \bigcup_{n=0}^{\infty} X^n,$$

where $X^1 = X$, $X^2 = X \times X$, $X^3 = X \times X \times X, ...$

For example, let $X = \{\mathbf{a}, \mathbf{b}, \mathbf{c}\}$. $\mathbf{aaa}$, $\mathbf{aba}$, $\mathbf{cba}$ are examples of elements of X^3, namely, strings of length 3. $X^0 = \{\varepsilon\}$ is the set of the empty string. We define for a scalar $a \in F$, $a\varepsilon = a$. We can construct a free vector space $\mathcal{V}_F \langle X^* \rangle$ over the field F generated by set X^*. We define the multiplication of two strings to be the juxtaposition (or concatenation) of the two strings. For example, **aba** multiplied by **cba** will be **abacba**. The strings are considered "associative". We stipulate the distributive laws regarding multiplication and addition. This way we obtain an associative algebra over F, called the free associative algebra generated by X, denoted by $\mathcal{A}_F \langle X^* \rangle$. It is an infinite dimensional algebra over F. This construction is used to construct tensor spaces.

As an example, let $X = \{x\}$ be a set of a single letter. The free algebra $\mathcal{A}_F \langle X^* \rangle$ is the same as the algebra of all the polynomials $F[x]$ in a single variable x in the form of $a_0 + a_1 x + a_2 x^2 + \cdots + a_n x^n$.

Appendix 2. Mathematical Structures

§1. Mathematical Structures

Modern mathematics is built on the foundations of set theory. The universe in a mathematical system is a set of elements, for example, the set of all real numbers, or the set of all points in the Euclidean space. A mathematical structure $(X, r_1, r_2, r_3, \ldots, r_k)$ is a set X, together with any number of relations $r_1, r_2, r_3, \ldots, r_k$. Each relation r_i could be an n-ary relation with a different n, meaning it could be a binary relation, a ternary relation, etc. Any or all of these relations could be mappings, because a mapping is a special case of relations. The set X is called the **universal set** (or **universe**, or **underlying set**) of the structure.

Oftentimes the universal set X is also called the **space** (or **abstract space**), and each element in the set is called a **point**. Note that these terms are borrowed from geometry but they are abstract now. For example, in the case of a complex Hilbert space, each point is a complex valued function.

A group $(G, +)$ is one example of mathematical structures. G is a set and $+$ is a mapping, or a binary operation $(+) : G \times G \to G$ in this case.

A partially ordered set $(S, \leq)$ is a structure, with S being a set and $\leq$ being a binary relation, which is a partial order.

Sometimes, a structure may have more than one underlying set involved. It is called a **many-sorted** system. We need to make clear each element is from which set in such a case. For example, a vector space over a field is a structure $(V, F, +, \cdot)$, where V is a set whose elements are called vectors and F is a set whose elements are called scalars. $(+) : V \times V \to V$ is a mapping called vector addition and $(\cdot) : F \times V \to V$ is a mapping called scalar-vector multiplication (the dot symbol is often omitted).

Usually, we stipulate that these relations $r_1, r_2, r_3, \ldots, r_k$ satisfy certain conditions. These conditions are called **axioms**. A mathematical structure with a set of axioms is called a **theory of the structure**. The same structure can be equipped with different axioms and hence to form different theories. One example is that Euclidean geometry and hyperbolic geometry have the same structure but different axioms. They are different theories of the same structure.

In this way, we have a clutter of hundreds of structures as branches of mathematics. For housekeeping's sake, it helps if we could sort these structures into categories.

Bourbaki divides the mathematical structures into three major types: algebraic structures, order structures and topological structures (this classification itself is heuristic in nature). They call these three types mother structures. I would like to suggest a modification to this scheme: we divide all the mathematical structures into two categories: **discrete structures** and **continuous structures**. Algebraic structures and order structures are examples of discrete structures. Topological structures and measure structures are examples of continuous structures. Geometry may have either or both of discrete and continuous structures, depending on how we study it. The study of geometries (Euclidean and hyperbolic) in the traditional way using incidence and metric relations ("colinear" and "congruence") following Euclid, Hilbert and Tarski, deals with discrete structures. However, the space in the geometry is considered a topological space. Differential geometry is an example in this respect, where we study continuous structures. Also in this sense, topology is considered a generalization of geometry.

Let us compare discrete structures and continuous structures. In a system of discrete structures $(X, r_1, r_2, r_3, \ldots, r_k)$, the relations r_i are defined on X, $X \times X$, $X \times X \times X$, etc. However, in a system of continuous structures, (X, r) for example, the relation or mapping is defined on the power set $\mathbf{P}(X)$ of X. The mappings are set functions. In the case of a topological space (X, τ), X is the universal set and τ is the topology, which can be viewed as a mapping $\tau : \mathbf{P}(X) \to \{0, 1\}$. For a subset $U \in \mathbf{P}(X)$, if $\tau(U) = 1$, then we say U is an open set. Namely, the topology τ can be viewed as the class of all open sets. Hence the mapping τ is a "set function" which assigns a real number to each subset of X (or a member of $\mathbf{P}(X)$).

Similarly, in a measure structure (X, μ), the mapping $\mu : \mathbf{M} \to \mathbb{R}$ is defined on $\mathbf{M} \subseteq \mathbf{P}(X)$, a family of subsets of X, called the family of measurable sets. Thus, the measure, μ is a "set function" which assigns a real number to each subset in the family of measurable sets. This is the reason why I propose to modify the three categories of structures of Bourbaki—a measure structure is a continuous structure in my sense but it is not a topological structure and it does not have a place in the three "mother structures" of Bourbaki.

It makes sense to think that continuous structures provide a means to describe the "congregation" of points, whether they are near to each other (as in topology), or how much "volume" they occupy (as in measure theory), by way of the set functions.

From the perspective of mathematical logic, continuous structures need higher order predicate logic to describe, than discrete structures because they involve subsets. Hence continuous structures are more complex than discrete structures. If a student finds the concepts in point-set topology are harder to understand than those (like incidence and congruence) in Euclidean geometry, this is part of the reason, namely they are indeed more complex. A "set function" (with the domain on a family of sets) is more complex than ordinary functions (with the domain on a single set). Of course, in set theory, a set of individual elements and a set of sets are not distinguished. It is the same concept, just set. However, if a family of sets gets too large, it no longer qualifies as a set, and it should be called a class and be expelled from the study of set theory. This is how we try to stay away from the paradoxes of naive set theory.

§2. Discrete Structures

Algebraic structures and order structures are two examples of discrete structures.

2.1 Algebraic Structures

Let X be the universal set. An algebraic structure $(X, \varphi_1, \varphi_2, \ldots, \varphi_k)$ usually has one or more mappings, $\varphi_1, \varphi_2, \ldots, \varphi_k$, most of the times binary operations, in place of those relations in $(X, r_1, r_2, \ldots, r_k)$. A mapping $\varphi : X \times X \to X$ is also called a binary operation. Examples of algebraic structures include groups, rings, fields, vector spaces, algebras over a field, tensor algebras, exterior algebras and geometric algebras.

2.2 Order Structures

An order structure is a structure $(X, \leq)$, where $\leq$ is a partial order. The order $\leq$ is in natural numbers as well as in real numbers and many other number systems. Number systems $\mathbb{N}$, $\mathbb{Z}$, $\mathbb{Q}$, and $\mathbb{R}$ all have total orders.

Some systems have a partial order. The partial order must satisfy certain axioms. These order structures include lattices and Boolean algebras.

§3. Continuous Structures

Topological structures and measure structures are two examples of continuous structures.

3.1 Topological Structures

The concept at the center of topology is the concept of continuity. The key concept to describe continuity is "near" or "neighborhood". A topological structure $(X, \mathbf{O})$ is a structure, in which a family $\mathbf{O}$ of "open sets" are specified. Using the concept of open set, all the familiar concepts we have encountered in analysis, like neighborhood, limit, continuous mapping, can be defined.

The topological structure can be viewed as a set function, namely the characteristic function of $\mathbf{O}$, $\tau : \mathbf{P}(X) \to \{0, 1\}$ on the power set $\mathbf{P}(X)$. A subset $U \in \mathbf{O}$ if and only if $\tau(U) = 1$.

3.2 Measure Structures

The concept of measure is the generalization of length, area and volume. In a measure space $(X, \mathbf{M}, \mu)$, X is a nonempty set as the universal set. $\mathbf{M} \subseteq \mathbf{P}(X)$ is a family of "good-natured" subsets of X. Here "good-natured" is in a different sense from that "good-natured" in the context of topology. $\mathbf{M}$ is the family of all measurable sets. $\mu : \mathbf{M} \to \mathbb{R}$ is a set function called the measure, which assigns a real number to each set in $\mathbf{M}$. The specification of the family $\mathbf{M}$ of measurable sets can also be defined by its characteristic function, which is a set function, $m : \mathbf{P}(X) \to \{0, 1\}$. A is a measurable set if and only if $m(A) = 1$.

Topological structures and measure structures are quite different. What they have in common is that these structures are defined on a family of subsets of X, instead of on X itself. That is the reason they are considered continuous structures. This makes sense with our intuition because the concept of continuity deals with congregation of elements of sets.

§4. Mixed Structures

Many systems have a mixture of multiple structures. Take the real numbers $\mathbb{R}$ for example. $\mathbb{R}$ has all these types of structures: algebraic structure, order structure, topological structure and measure structure. $\mathbb{R}$ has two binary operations, addition and multiplication. $\mathbb{R}$ has an order structure with a total order $\leq$. $\mathbb{R}$ also has a topological structure, as a complete metric space, with open sets being arbitrary union of open intervals. Furthermore, Lebesgue measure as the measure structure, which is the generalization of length, is defined on $\mathbb{R}$. From the structure point of view, $\mathbb{R}$ can be characterized as a "complete Archimedean ordered field".

> *** Computer Science**: Connection to object-oriented programming
>
> The idea of structures was reinvented in computer science by programmers in 1970s, in the context of abstract data types and object-oriented programming. Object-oriented programming had become a dominating programming paradigm by 1990s. However, because of the separation of disciplines, mathematicians are not trained for programming and programmers are not trained for abstract mathematics. Hardly anyone realized the connection between the "objected-oriented" programming and mathematical structures. In fact, objected-oriented programming has everything in the idea of mathematical structures, except the obscure and awkward name. When a class of objects is defined, a set of function templates (which are alternatively called methods) are also defined. These

function templates specify what are the input domain and output domain for each function, just like the mappings in the mathematical structures. From the perspective of mathematics, "structured programming" could be a better name for "object-oriented" programming. However, in the computer science jargon, "structured programming" is referred to an older programming paradigm which simply means "not object-oriented". Historically, "structured programming" was so named to distinguish from an even older programming paradigm known as the "spaghetti code", which was considered "unstructured".

Appendix 3. Axiomatic Systems

§1. Undefined Concepts and Axioms

Rigor is the heart of mathematics. All the terms used in mathematics should be precisely defined. However, to define a new term, we need to use old terms. To define the old terms, we need to use even older terms.

Look at the following conversation between two friends A and B. B is blind while A has normal vision.

A: Shall we stop by that shop and drink a glass of milk?

B: What is milk?

A: Milk is a white liquid.

B: Liquid I know, but what is white?

A: White is the color of the feathers of a swan.

B: Feathers I know, but what is a swan?

A: A swan is a bird with a crooked neck.

B: Neck I know, but what is crooked?

A becomes impatient. He grabs B's arm and holds it straight.

A: Look! This is straight.

A then bends B's arm.

A: And this, is crooked!

> B: Oh! Now I understand milk. Let's go and have a glass of milk!

This was a story in a book by Max Born. It was a French friend instead of a blind man in his original version. Some modified version of the story later became an apocryphal anecdote often mistakenly attributed to Albert Einstein (a phenomenon known as the Matthew effect[11]).

I find this story a good example to illustrate the idea of axiomatic systems. Look at the chain to define the concept of milk: milk ← liquid, white ← feather, swan ← bird, crooked … This chain of definitions cannot go on forever or become circular. We have to cut the chain somewhere and be contended with certain terms undefined, like in this case, "crooked". The term "crooked" is only explained by physical intuition of bending the arm, rather than defined rigorously with more terms.

***

*** Historical Note**: The story as in Max Born's book

Max Born published a book in 1935, *The Restless Universe* (translated by Winifred M. Deans, Blackie and Son, London, p. 75). In his story, it is a French friend, instead of a blind friend. The message that this story conveys is possibly that explaining relativity theory to a layperson is as difficult as explaining the white color to a French.

A friend of mine was once at a dinner-party and the lady next to him said: "Professor, do tell me in a few words what this theory of relativity really is." He replied: "of course I will—provided you will let me tell you this little story first. I was going for a walk with a French friend and we got thirsty. By and by we came to a farm and I said: 'Let's buy a glass of milk here.' 'What's milk?' 'Oh, you don't know what milk is? It's the while liquid that—' 'What's white?' 'White? you don't know what that is either? Well, the swan—' 'What's swan?' 'Swan, the big bird with the bent neck.' 'What's bent?' 'Bent? Good heavens, don't you know that? Here, look at my arm: when I put it so, it's bent!' 'Oh, that's bent, is it? Now I know what milk is!'" Perhaps, like the lady, you do not want to hear any more about relativity.

***

---

[11]Such examples abound, like Kronecker was not the first to define Kronecker product; Kolmogorov was not the first to define Kolmogorov complexity.

We cannot use circular definitions either, like the following example:

Definition: The radius of a circle is one half of the diameter.

Definition: The diameter of a circle is two times the radius.

The following are a few definitions from Euclid's *Elements*:

A point is that which has no part.

A line is breadthless length.

A straight line is a line which lies evenly with the points on itself.

A surface is that which has length and breadth only.

A plane surface is a surface which lies evenly with the straight lines on itself.

Obviously these are not good definitions without *part, breadthless length,* and *lies evenly* being defined. In mathematics, we have to live with the fact that some terms cannot be defined. These terms are called **undefined terms**, or **primitive terms**. All other terms can be defined using these primitive terms. In the analogy of the milk story above, "crooked" is an undefined term, and "milk" is defined using "crooked".

The similar reasoning applies to the proof process. To prove a theorem, we need to use the old proved theorems, and to prove those old theorems, we need even older theorems. This process cannot go on forever and we have to stop somewhere, where we select a group of statements and we assume they are true. These hypotheses are called **axioms**.

The ancient view and the modern view of axioms are different. In the ancient times, people thought axioms were self evident truth about the world, which did not need to be proved, like the postulates in the *Elements*. In modern times, these axioms are viewed as arbitrary assumptions. It was an interesting new chapter in mathematics history that Lobachevsky and Bolyai negated one axiom, the axiom of parallels in the Euclidean geometry and created a new geometry—hyperbolic geometry.

In the view of formalists, represented by David Hilbert, those terms are just meaningless symbols. According to Hilbert, it would be the same if we replace "points", "lines" and "planes" by "tables", "chairs" and "beer-mugs". When we associate these abstract symbols with meanings, we have an **interpretation**, or a **model** of the system.

§2. Axiomatic Systems—From Ancient to Modern Times

The standard for rigor in mathematics has also been evolving through history. The *Elements* of Euclid had been regarded as the standard of rigor for two thousand years until the late 1800s when it was criticized and improved. One drawback of Euclid is that he tries to define every concept, like point and line, without a clear declaration of the undefined primitive concepts. Isaac Newton tried to follow Euclid's approach in his *Principia*. Hence his *Principia* suffers the same drawback. Benedict de Spinoza (1632–1677) even mimicked the style of the *Elements* in his work *Ethics*. He started with definitions, axioms and postulates about God and human mind, etc., and proceeded to prove propositions and corollaries.

*** Excerpts from Spinoza's *Ethics***

Part II. On the Nature and Origin of the Mind

AXIOMS

(1) The essence of man does not involve necessary existence; that is to say, the existence as well as the non-existence of this or that man may or may not follow from the order of nature.

(2) Man thinks.

(3) Modes of thought, such as love, desire, or the emotions of the mind, by whatever name they may be called, do not exist unless in the same individual exists the idea of a thing loved, desired, etc. But the idea may exist although no other mode of thinking exist.

(4) We perceive that a certain body is affected in many ways.

(5) No individual things are felt or perceived by us except bodies and modes of thought.

POSTULATES

(1) The human body is composed of a number of individual parts of diverse nature, each one of which is composite to a high degree.

(2) Of the individual parts of which the human body is composed, some are fluid, some soft, and some hard.

(3) The individual parts composing the human body, and consequently the human body itself, are affected by external bodies in many ways.

(4) The human body needs for its preservation many other bodies by which it is, as it were, continually regenerated.

(5) When a fluid part of the human body is determined by an external body, so that it often strikes upon another which is soft, the fluid part changes the plane of the soft part and leaves upon it, as it were, some traces of the impelling external body.

(6) The human body can move and arrange external bodies in many ways.

Proposition 39. If a man hates another, he will endeavor to do him evil unless he fears a greater evil will therefrom arise to himself and, on the other hand, he who loves another will endeavor to do him good by the same rule.

Proof. To hate a person (Note, Prop. 13, pt. 3) is to imagine him as a cause of sorrow, and therefore (Prop. 28, pt. 3) he who hates another will endeavor to remove or destroy him. But if he fears lest a greater grief or, which is the same thing, a greater evil should fall upon himself, and one which he thinks he can avoid by refraining from inflicting the evil he meditated, he will desire not to do it (Prop. 28, pt. 3); and this desire will be stronger than the former with which he was possessed of inflicting the evil, and will prevail over it (Prop. 37, pt. 3). This is the first part of the proposition. The second is demonstrated in the same way. Therefore if a man hates another, etc. —Q.E.D.

. . .

Proposition 43. Hatred is increased through return of hatred, but may be destroyed by love.

Proof. If we imagine that the person we hate is affected with hatred toward us, a new hatred is thereby produced (Prop. 40, pt. 3), the old hatred still remaining (by hypothesis). If, on the other hand, we imagine him to be affected with love toward us, in so far as we imagine it (Prop. 30, pt. 3) shall we look upon ourselves with joy and endeavor (Prop. 29, pt. 3) to please him, that is to say (Prop. 41, pt. 3), in so far shall we endeavor not to hate him nor to affect him with sorrow. This effort (Prop. 37, pt. 3) will be greater or less as the emotion from which it arises is greater or less, and, therefore, should it be greater than that which springs from hatred, and by which (Prop. 26, pt. 3) we endeavor to affect with sorrow the object we hate, then it will prevail and banish hatred from the mind. —Q.E.D.

...

Part III. On the Origin and Nature of the Emotions

Corollary 2. (of Prop. 55) No one envies the virtue of a person who is not his equal.

Proof. Envy is nothing but hatred (Note, Prop. 24, pt. 3), that is to say (Note, Prop. 13, pt. 3), sorrow, or in the other words (Note, Prop. 11, pt. 3), a modification by which the effort of a man or his power of action is restrained. But (Note, Prop. 9, pt. 3) a man neither endeavors to do nor desires anything except what can follow from his given nature, therefore a man will not desire to affirm of himself any power of action or, which is the same thing, any virtue which is peculiar to another nature and foreign to his own. His desire, therefore, cannot be restrained, that is to say (Note, Prop. 11, pt. 3), he cannot feel any sorrow because he contemplates a virtue in another person altogether unlike himself, and, consequently, he cannot envy that person, but will only envy one who is his own equal, and who is supposed to possess the same nature. —Q.E.D.

In early 1900s, there were some attempts to axiomatize some branches of physics, like thermodynamics [Caratheodory (1909)] and the theory of relativity [Reichenbach (1924)]. Even though they list some statements as axioms, those statements are vague. They do not even disclose a list of undefined/primitive concepts, but rather use vague terms. These efforts of axiomatization of physics are not more rigorous than the *Ethics* of Spinoza.

In late 1800s and early 1900s, some authors attempted to give a more rigorous axiomatic system for the Euclidean geometry. These include Pasch, Peano, Pieri, Padua, Veronese and Hilbert. In Hilbert's system, the undefined concepts are "point", "line", "plane", "between" and "congruent". Hilbert then poses twenty axioms in five groups.

*** Axioms of Hilbert**

Axiom Group I: Axioms of Incidence

(1) For every two points A, B there exists a line a that contains each of the points A, B.
(2) For every two points A, B there exists no more than one line that contains each of the points A, B.
(3) There exist at least two points on a line. There exist at least three points that do not lie on a line.

(4) For any three points A, B, C that do not lie on the same line there exists a plane α that contains each of the points A, B, C. For every plane there exists a point which it contains.

(5) For any three points A, B, C that do not lie on one and the same line there exists no more than one plane that contains each of the three points A, B, C.

(6) If two points A, B of a line a lie in a plane α then every point of a lies in the plane α.

(7) If two planes α, β have a point A in common then they have at least one more point B in common.

(8) There exist at least four points which do not lie in a plane.

Axiom Group II: Axioms of Order

(1) If a point B lies between a point A and a point C then the points A, B, C are three distinct points of a line, and B then also lies between C and A.

(2) For two points A and C, there always exists at least one point B on the line AC such that C lies between A and B.

(3) Of any three points on a line there exists no more than one that lies between the other two.

(4) Let A, B, C be three points that do not lie on a line and let a be a line in the plane ABC which does not meet any of the points A, B, C. If the line a passes through a point of the segment AB, it also passes through a point of the segment AC, or through a point of the segment BC. (Expressed intuitively, if a line enters the interior of a triangle, it also leaves it.)

Axiom Group III: Axioms of Congruence

(1) If A, B are two points on a line a, and A' is a point on the same or on another line a' then it is always possible to find a point B' on a given side of the line a' through A' such that the segment AB is congruent or equal to the segment $A'B'$. In symbols, $AB \equiv A'B'$.

(2) If a segment $A'B'$ and a segment $A''B''$ are congruent to the same segment AB, then the segment $A'B'$ is also congruent to the segment $A''B''$, or briefly, if two segments are congruent to a third one they are congruent to each other.

(3) On the line a let AB and BC be two segments which except for B have no point in common. Furthermore, on the same or on another line a' let $A'B'$ and $B'C'$ be two segments which except for B' also have no point in common. In that case, if $AB \equiv A'B'$ and $BC \equiv B'C'$ then $AC \equiv A'C'$.

(4) Let $\angle(h,k)$ be an angle in a plane α and a' a line in a plane α' and let a definite side of a' in α' be given. Let h' be a ray on the line a' that emanates from the point O'. Then there exists in the plane α' one and only one ray k' such that the angle $\angle(h,k)$ is congruent or equal to the angle $\angle(h',k')$ and at the same time all interior points of the angle $\angle(h',k')$ lie on the given side of a'. Symbolically, $\angle(h,k) \equiv \angle(h',k')$. Every angle is congruent to itself, i.e., $\angle(h,k) \equiv \angle(h,k)$ is always true.

(5) If for two triangles ABC and $A'B'C'$ the congruences $AB \equiv A'B'$, $AC \equiv A'C'$, $\angle BAC \equiv \angle B'A'C'$ hold, then the congruence $\angle ABC \equiv \angle A'B'C'$ is also satisfied.

Axiom Group IV: Axioms of Parallels

(1) (Euclid's Axiom) Let a be any line and A a point not on it. Then there is at most one line in the plane, determined by a and A, that passes through A and does not intersect a.

Axiom Group V: Axioms of Continuity

(1) (Axiom of measure or Archimedes' Axiom) If AB and CD are any segments then there exists a number n such that n segments CD constructed contiguously from A, along the ray from A through B, will pass beyond the point B.

(2) (Axiom of line completeness) An extension of a set of points on a line with its order and congruence relations that would preserve the relations existing among the original elements as well as the fundamental properties of line order and congruence that follows from Axioms I–III, and from V (1) is impossible.

The rigor has been improved in Hilbert's treatment, but still not perfect in today's standard. Hilbert, as well as his contemporaries, used the natural language, but today we use symbolic languages (or formal languages), because natural languages are vague.

Let us be confined to the first order theories. A first order theory consists of a first order language and a set of axioms. Any statement in the theory is a sentence (string of symbols) complying with a grammar. First we need to define the alphabet (the set of symbols) and the grammar. Any legitimate statement (regardless true or false) complying with the grammar is called a well-formed formula (wff). The symbols in the alphabet consist of [Wolf (2005)]:

(1) a denumerable list of variables $v_0, v_1, v_2, \ldots$;
(2) for each natural number n, a set of n-ary relation symbols (also called predicate symbols);
(3) the equality symbol $=$;
(4) the logic connectives $\vee, \wedge, \sim, \rightarrow, \leftrightarrow$;
(5) the quantifiers $\forall, \exists$;
(6) parentheses and the comma.

(There can also be constant symbols and function symbols but these can be viewed as special cases.) The first two types of symbols correspond to the traditional "undefined concepts", but we distinguish the "undefined entities" from "undefined predicates (or relations)". The entities could be divided into multiple categories, making the system **many-sorted**. Now look at Hilbert's system of geometry, it is many-sorted with three categories "point", "line" and "plane". The concepts "between" and "congruent" are really undefined relations. There are some problems with Hilbert's systems:

First, "congruent" is overloaded. It actually represents two different relations: congruence of two line segments and congruence of two angles.

Second, neither line segment nor angle is one of the primitive concepts.

Third, the axioms of continuity are second order predicates in nature.

We shall illustrate what a modern axiomatic system is like with Tarski's first order system of elementary plane geometry [Tarski (1959)], which he called $\mathcal{E}_2$. Compare Tarski's system with Euclid's *Elements*, Spinoza's *Ethics* (1677), Newton's *Principia* (1687), Hilbert's *Foundations of Geometry* [(1899)], Caratheodory's axiomatization of thermodynamics [(1909)] and Reichenbach's axiomatization of relativity [(1924)], you will see the differences.

Tarski's system is single-sorted. The only undefined entity is "point". He uses two undefined/primitive relations, a ternary relation *betweenness* $\beta(x, y, z)$ and a quaternary relation *equidistance* $\delta(x, y, z, w)$. We can use $xy \equiv zw$ as a shorthand for $\delta(x, y, z, w)$.

* Axioms of Tarski's $\mathcal{E}_2$

Note that for simplicity, some universal quantifiers are omitted. For instance, $\beta(xyx) \rightarrow (x = y)$ means $\forall xy \beta(xyx) \rightarrow (x = y)$.

(1) Identity of betweenness
$\beta(xyx) \rightarrow (x = y)$.

(2) Transitivity of betweenness
$\beta(xyu) \wedge \beta(yzu) \rightarrow \beta(xyz)$.

(3) Connectivity of betweenness
$\beta(xyz) \wedge \beta(xyu) \wedge (x \neq y) \rightarrow \beta(xzu) \vee \beta(xuz)$.

(4) Reflexivity of equidistance
$xy \equiv yx$.

(5) Identity of equidistance
$xy \equiv zz \rightarrow (x = y)$.

(6) Transitivity of equidistance
$xy \equiv zu \wedge xy \equiv vw \rightarrow zu \equiv vw$.

(7) Pasch's axiom
$\beta(xtu) \wedge \beta(yuz) \rightarrow \exists v \beta(xvy) \wedge \beta(ztv)$.

(8) Euclid's axiom
$\beta(xut) \wedge \beta(yuz) \wedge (x \neq u) \rightarrow \exists vw \beta(xzv) \wedge \beta(xyw) \wedge \beta(vtw)$.

(9) Five-segment axiom
$xy \equiv x'y' \wedge yz \equiv y'z' \wedge xu \equiv x'u' \wedge yu \equiv y'u' \wedge \beta(xyz) \wedge \beta(x'y'z') \wedge (x \neq y) \rightarrow zu \equiv z'u'$.

(10) Axiom of segment construction
$\exists z \beta(xyz) \wedge yz \equiv uv$.

(11) Lower dimension axiom
$\exists xyz[\sim \beta(xyz) \wedge \sim \beta(yzx) \wedge \sim \beta(zxy)]$.

(12) Upper dimension axiom
$xu \equiv xv \wedge yu \equiv yv \wedge zu \equiv zv \wedge (u \neq v) \rightarrow \beta(xyz) \vee \beta(yzx) \vee \beta(zxy)$.

(13) Axiom schema of continuity
All sentences of the form
$\forall vw \ldots \{\exists z \forall xy[\varphi \wedge \psi \rightarrow \beta(zxy)] \rightarrow \exists u \forall xy[\varphi \wedge \psi \rightarrow \beta(xuy)]\}$, where φ stands for any formula in which the variables $x, v, w, \ldots$, but neither y nor z nor u, occur free, and similarly for ψ, with x and y interchanged.

* Axioms of ZF Set Theory

The language is single-sorted. Every object is a set. There is only one binary relation symbol $\in$.

(1) Extensionality
$\forall xy[x = y \leftrightarrow \forall u(u \in x \leftrightarrow u \in y)]$.

(2) Pairing
$\forall xy \exists z \forall u(u \in z \leftrightarrow u = x \vee u = y)$.

(3) Union
$\forall x \exists y \forall u(u \in y \leftrightarrow \exists w \in x(u \in w))$.

(4) Empty set
$\exists x \forall y \sim (y \in x)$.

(5) Infinity
$\exists x[\emptyset \in x \wedge \forall y \in x((y \cup \{y\}) \in x]$.
Note some symbols here are not primitive symbols but rather defined symbols: $\emptyset$ is a shorthand for the empty set; $\cup$ is a shorthand for union and $\{y\}$ is a shorthand for the singleton set with one element y.

(6) Power set
$\forall x \exists y \forall u(u \in y \leftrightarrow u \subseteq x)$.
Note $\subseteq$ is a shorthand for subset.

(7) Replacement schema
$[\forall x \in a \exists! y \mathrm{P}(x, y)] \to [\exists b \forall y(y \in b \leftrightarrow \exists x \in a \mathrm{P}(x, y))]$,
where $\exists!$ means "exists unique", and $\mathrm{P}(x, y)$ is a formula that does not contain b as a free variable.

(8) Regularity or Foundation
$\forall x[x \neq \emptyset \to \exists y \in x(x \cap y = \emptyset)]$,
where $\cap$ is a shorthand for intersection.

* Axiom of Choice

$[\forall u \in x(u \neq \emptyset) \wedge \forall uv \in x(u \neq v \to u \cap v = \emptyset)]$
$\to \exists y \forall u \in x \exists! w \in u(w \in y)$.

*** Mathematical Logic**: First order, second order and higher order predicates

In a first order language discussed above, the universal quantifier $\forall$ quantifies over the domain of a universal set. For instance, $\forall xy \, \delta(xyyx)$ means "for all objects x and y in the domain, the relation $\delta(xyyx)$ holds". The objects are the "points" and δ is one of the two primitive predicates (or relations). The variables like x and y are not allowed to take values from arbitrary predicates. That is to say, we are not allowed to use predicate variables, for instance, $\forall \varphi(\cdots)$, to say "for all predicate φ ...". This is why this language is called the first order predicate language.

If we do allow to quantify over (first order) predicates, we have a second order predicate language. If we allow to quantify over second order predicates, we have a third order predicate language. Similarly we can have even higher order predicates. Why do we distinguish the different orders of the predicates? This is to avoid the liar paradox, a paradox in Ancient Greece: a person who is a Cretan makes a statement: "All Cretans are liars". Its modern version is the following statement A:

A: This statement (A) is false.

Is A true or false? Assume A is true, then read what A says—A is false. Assume A is false, then read what A says—"A is false". This means this statement is telling the truth. Therefore, A is true.

What is the cause that leads to this paradox? Look at the following statements:

B: $\forall xy \, \delta(xyyx)$.

C: The statement B is true.

The statement C is a statement about the truth of another statement. If statement B is in one language, the statement C is in a meta-language, at a higher level. The cause of the liar paradox is to mix a language with a meta-language, or mix languages at different levels. Inside statement A, it talks about the truth of the statements, especially the truth value of itself.

A predicate corresponds to a set (or subset). Any statement involving quantifying over subsets is second order in nature. Tarski's $\mathscr{E}_2$ is a first order system. So we are not allowed to quantify over subsets. That is why Axiom 13 in Tarski's system is an axiom schema, meaning infinitely many first order axioms in that form.

We see the formal symbolic language with a grammar is essential to the rigor. In the following, we look at some statements in the context of Tarski's elementary plane geometry $\mathscr{E}_2$.

Look at this statement: $\forall xyz[xy \equiv yz \wedge yz \equiv zx]$. It translates to *every triangle is an equilateral triangle*. It is false, but it is still a legitimate wff. However, look at the following string of symbols:

9ñz\$7*û3bpÄ<@§.

Is it true or false? This is simply nonsense. It contains symbols out of our alphabet (this is actually a chunk of characters of an image file when opened with a text editor). The alphabet of a formal language is the counterpart of the vocabulary of a natural language.

Look at the next string of symbols:

$\forall xy \exists zu[\beta(xy) \wedge xy \equiv zu \wedge f(xy)]$.

This is not a wff either. β is a ternary relation while here it is applied to two variables, and f is not one of the "primitive relations". So basically this is also a nonsense, and it has no meaning. We cannot talk about whether it is true or false. Anything which is not a wff complying with the formal grammar is gibberish nonsense. One of the advantages of a formal symbolic language is this: to decide whether something is a meaningful statement (wff) or gibberish nonsense, we can rely on a set of objective rules, rather than rely on philosophical debates. So we establish a rigorous definition of and objective test for "nonsense". A wff with a false value is not nonsense, but anything is not a wff is simply nonsense and is automatically disqualified for serious discussion in our theory. If we do not have a rigorous formal language, people may keep asking nonsense questions and making nonsense statements, like "Why do we remember the past but we do not remember the future?" or "Why does time have to go in one direction and cannot go backward?"[12]

The problem with some of the attempted "axiomatic systems" using natural languages is vagueness. There is something worse than being wrong in this world. That is being vague. A vague statement is like gibberish. It does not have a clear meaning, and we cannot talk about whether it is true or false.

The axiomatization of physics is the 6th of the 23 famous open problems that Hilbert presented in 1900. There have been some attempts but this problem has not been solved satisfactorily so far. Merely selecting a

---

[12]The statement versions are: we remember the past and we do not remember the future; time goes in one direction; time cannot go backward. If we do not understand what time is, we certainly will not understand it better by talking nonsense about it.

few vague statements and labeling them as Axiom 1, Axiom 2, ..., like Spinoza's axiom "Man thinks", does not qualify the work as an axiomatic system in the modern sense. These attempts include the treatment of thermodynamics by C. Caratheodory [(1909)] and the treatment of the theory of relativity by H. Reichenbach [(1924)], which we have mentioned earlier. They use natural languages and vague terms freely. They did not specify the primitive terms. These treatments are not more rigorous than the "axiomatic treatment" by Spinoza about God and human nature in his *Ethics*.

The theory of special relativity in Einstein's 1905 paper [Einstein (1905)] is not an axiomatic system, although Einstein started his theory with two "postulates"—the principle of relativity and the principle of constancy of the velocity of light. Neither is his general theory of relativity [Einstein (1916)]. In 1950s, there was a heated debate over the twins paradox between Herbert Dingle on one side and more than a dozen other physicists on the opposite side. Dingle first argued that the theory of special relativity implies equal aging, but later he switched to a position that special relativity has a contradiction and hence is inconsistent. The problem is, the twins paradox is not rigorously formulated in a rigorous formal sentence. Furthermore, when the theory is not properly and rigorously axiomatized, we don't have a clear agreement on what the theory of special relativity is exactly. If we don't have a rigorous language, we are not even sure that the "twins paradox" is formulated as a valid sentence (wff) within the formal language of special relativity. It is still an unsettled debate today among physicists whether motion in an accelerated (non-inertial) reference frame without gravity belongs to the domain of special relativity or not. Without a satisfactory axiomatization, we cannot discuss whether the theory is consistent or not either. It is meaningless to debate about a vaguely formulated statement in a vaguely formulated theory using a vague language. If the theory of relativity is properly axiomatized, it is possible to prove its relative consistency using the method of models.

If we keep the fundamental terms undefined, may different people have different understanding of those terms? Will this cause disagreement and debate? Will mathematics be built on the foundations of sand?

For instance, what is a straight line? It is undefined. What if for a line, some people think it is straight while others think it is curved? Do we debate and fight like politicians do?

We do have tests to determine whether a line is straight or not. We use the axioms. A straight line should behave in a way that is stipulated by the axioms.

This leads to the modern view of axioms. They can be regarded as the definitions of those undefined terms. That is, the axioms are just "disguised definitions". This view is often attributed to Henri Poincaré [(1899)], but it was expounded much earlier by José Gergonne [(1818)]. Gergonne regarded the axioms as "implicit definitions".

In Euclidean geometry, the axiom of parallels says that in a plane, passing through a point P not lying on a line l, there is at most one line parallel to l. However, in hyperbolic geometry, the axiom of parallels claims there exist at least two such lines passing through P and not intersecting l. Hyperbolic geometry is a sound geometry just as Euclidean geometry is sound. Now we find a reason to explain this. It is simply because that the term of straight line is undefined. The axioms of Euclidean geometry and the axioms of hyperbolic geometry are basically two different definitions of straight lines. The hyperbolic straight lines in hyperbolic geometry are simply curves in the Euclidean sense. This is demonstrated in the many models, like Poincaré disk model, or Gans model. It is also true that the Euclidean straight lines are curves according to the axiomatic definition of straight lines in hyperbolic geometry.

This is the reason, for example, in group theory, the statements about those operations in a group, like the associative law, etc., are called axioms of groups in some books, while they are called the definition of groups in some other books.

§3. Consistency, Independence and Completeness

3.1 Consistency

A set of axioms is consistent if no contradiction can be reached as a logical consequence of these axioms. Now that the axioms have lost their value as the absolute truth, consistency seems to be a minimal requirement.

How do we demonstrate, or prove that a set of axioms are consistent? We use models. We can show an axiomatic system is consistent if we can find one model that satisfies the set of all axioms. Here we are talking about relative consistency. Take plane Euclidean geometry for example. In the analytic geometry, a point in the plane *is represented* by an ordered pair of real numbers (x, y) as coordinates. A straight line *is represented* by the set of all ordered pairs of real numbers (x, y) that satisfy an equation $ax + by + c = 0$. This is a practical technique invented by Descartes and

Fermat. However, when Hilbert studies the axioms of Euclidean geometry, he states a point *is* (or *is interpreted* to be) an ordered pair of real numbers (x, y), and a straight line *is* (or *is interpreted* to be) the set of all ordered pairs of real numbers (x, y) that satisfy an equation $ax + by + c = 0$. With this interpretation (or model), he demonstrates, the points and straight lines so defined (or so interpreted) satisfy all the axioms of Euclidean geometry. Hence if there is any contradiction in Euclidean geometry, this contradiction can be translated to a contradiction in the system of real numbers. Therefore, if we trust that the system of real numbers is consistent, then Euclidean geometry is consistent.

Eugenio Beltrami was the first to give models of non-Euclidean geometry and hence settled the consistency problem of non-Euclidean geometry (and the independence of Euclid's axiom of parallels). It is demonstrated that if Euclidean geometry is consistent, so is hyperbolic non-Euclidean geometry.

All the models are axiomatic systems in mathematics. The consistency proved using these models is only relative consistency: if the system of real numbers is consistent, then Euclidean geometry is consistent; if Euclidean geometry is consistent, then hyperbolic geometry is consistent. There is no way we can prove absolute consistency. I would introduce a concept of physical models. A physical model is a model that we find in the material world for an axiomatic system. All the primitive concepts, entities and relations should be interpreted in the material world. This goes out of the mathematical world. What can a physical model do? First, it renders an application of the mathematical theory in the real world. Second, it gives us some confidence of its consistency. I would call it physical consistency. For example, for Euclidean geometry, we may find two physical models. In one model, we interpret a straight line (segment) as a taut string. For an alternative model, we may interpret a straight line as a light path. In small distance scales, both models work well. There are difficulties to validate either model in the large scale. We cannot extend a taut string from the earth to Mars, not even to the moon.

The idea of proving relative consistency using models (within mathematical systems) was fairly new (Beltrami, 1868). Before then, the consistency of Euclidean geometry was never questioned. For two thousand years, people believed the Euclidean geometry was the absolute truth. Why? Because it has physical models and Euclidean geometry was regarded as the truth about our physical world. According to an anecdote, Gauss measured the angle sum of a big triangle formed by mountain peaks in Germany to test the truth of Euclidean geometry. Is this a good way to validate the truth in geometry? Put aside the fact the deviation of the angle sum from π is too small compared with experimental error, suppose indeed Gauss finds the angle sum is significantly less than π, what does this prove or disprove? The scale with these mountain peaks is too big and it is not practical to measure them using taut strings, not to mention long rods. Gauss must have measured with surveying tools, making use of light. So even if he finds a discrepancy from π, it only shows that the light path is not a good model for the straight line in Euclidean geometry. It may show that the light path might be a good model for hyperbolic straight lines.

3.2 Independence

If we say one axiom is independent of the other axioms in an axiomatic system, we mean that this axiom is not a logical consequence of other axioms. It is not absolutely necessary to require that all the axioms are independent in an axiomatic system. Requiring each axiom to be independent of others can keep the system minimal. Another advantage of independent axioms is that we can see more clearly each theorem is the consequence of what axioms. Sometimes, having the redundant axioms in the system is for more convenience, or better symmetric looking for mnemonic purposes.

To demonstrate that one axiom A is independent of other axioms in the system, it suffices to show that both A and $\sim$ A are consistent with other axioms in the system, and again we demonstrate this using models.

In the history of geometry, mathematicians conjectured that Euclid's fifth axiom, the axiom of parallels, could be a consequence of other axioms, because it is lengthier and does not look like simple and evident truth while other axioms do. Great efforts of many great minds were put into the endeavor of proving this conjecture. In the end, it was shown instead that the axiom of parallels is independent of other axioms.

Lobachevsky and Bolyai are often credited for creating non-Euclidean geometry, or hyperbolic geometry. Hyperbolic geometry is also believed to have been known to Gauss through his private correspondence, although he did not publish anything on this because of his fear of the "outcry of the Boeotians".

E. Beltrami gave the first consistency proofs of hyperbolic geometry (which actually proved the independence of Euclid's axiom of parallels at the same time) in 1868 with an interpretation (or model) on a pseudosphere. His pivotal contributions are often less known and underappreciated by the general public. Even about the contributions of the consistency proof using models, his name is often overshadowed by names like Poincaré or Klein. In the popular folklore of the history of non-Euclidean geometry, Lobachevsky and Bolyai were the heroes with revolutionary ideas. Their contemporaries were all mediocre and had a tendency of resisting revolutionary ideas. This is rather a misconception. If either Lobachevsky or Bolyai had shown a consistency proof (which was to be shown later by Beltrami), people would have accepted the new geometry immediately without resistance. If Gauss had known a proof for the consistency of hyperbolic geometry, he wouldn't have feared the Boeotians to get his result published. Beltrami's idea and method were no less revolutionary than Lobachevsky and Bolyai.

Confucius said: "If one gets to know the truth in the morning, he can die content in the evening." Hilbert said: "We must know. We shall know." Confucius' words were written in the book *The Analects*, while Hilbert's words were engraved on his tombstone.

Neither Lobachevsky nor Bolyai died content. Lobachevsky developed a faith that his new geometry was consistent, because the trigonometric formulas he discovered in his new geometry bore a close resemblance to those in the Euclidean spherical geometry. If we replace the trigonometric functions $\sin\theta$, $\cos\theta$ and $\tan\theta$ in spherical geometry by the hyperbolic functions $\sinh\theta$, $\cosh\theta$ and $\tanh\theta$, we obtain the formulas for hyperbolic geometry (this corresponds to replacing $e^{i\theta}$ by e^{θ}), but that is only some heuristic rather than a proof. Lobachevsky published papers and books promoting his new geometry repeatedly in different languages—Russian, German and French, but did not receive any attention or recognition in his lifetime. His life was filled with frustration and in his last years, he suffered from blindness and financial difficulties. For Bolyai, although he "proudly published his non-Euclidean geometry, there is evidence that he doubted its consistency because in papers found after his death he continued to try to prove the Euclidean parallel axiom" [Kline (1972), p. 914]. The lives of Lobachevsky (died 1856), Bolyai (died 1860) and Gauss (died 1855) were just a little too short to see the truth: the consistency proof by Beltrami in 1868. If only they could have lived a little longer ... and died content.

3.3 Completeness

We say an axiomatic system is complete, if any statement can be proved, or disproved in the system. Some axiomatic systems are complete. For instance, the first order elementary Euclidean plane geometry $\mathscr{E}_2$ of Tarski is complete, which is proved by Tarski. Some other systems are incomplete, like groups and the first order Peano system of natural numbers. Completeness is not an absolute requirement for axiomatic systems either. For example, the axioms of groups is not complete and this is just fine, because we have a variety of groups that are not isomorphic to each other. However, for the Peano system of natural numbers, mathematicians including Hilbert, the advocate of axiomatic methods, rather wished it to be complete to uniquely characterize the natural number system. It was to most mathematicians' dismay and disappointment that Kurt Gödel proved that the first order Peano system of natural numbers is incomplete and it cannot be made complete by adding more axioms to it.

That an axiomatic system is incomplete means that there are statements independent of those axioms. Recall that the axioms in a system can be viewed as implicit definitions of the primitive terms. So in an incomplete axiomatic system the definitions are incomplete, or not precise enough to have all the questions regarding the primitive terms to have a definitive "yes" or "no" answer. P. Cohen proved that Cantor's continuum hypothesis is independent in the ZFC set theory. Cohen's explanation of this independence result is that "The notion of a set is too vague for the continuum hypothesis to have a positive or negative answer." Take geometry for another example. Absolute geometry is Euclidean geometry with the axiom of parallels removed. Euclid's axiom of parallels is independent of the axioms of absolute geometry. If we take the axioms in absolute geometry as the definition of straight lines, we must say the concept of a straight line as defined by the axioms of absolute geometry is too vague for some questions regarding straight lines to have a definitive answer.

Bibliography

Arnold, V. I. (1997). *Mathematical Methods of Classical Mechanics*, 2nd ed. (Springer).

Auer, P., Cesa-Bianchi, N. and Fisher, P. (2002). Finite-time Analysis of the Multiarmed Bandit Problem, *Machine Learning*, **47**(2-3), pp. 235–256.

Beem, J. K., Ehrlich, P. E. and Easley, K. L. (1996). *Global Lorentzian Geometry*, 2nd ed. (Marcel Dekker).

Bishop, R. and Goldberg, S. (1980). *Tensor Analysis on Manifolds* (Dover).

Blanuša, D. (1955). Über die Einbettung Hyperbolischer Räume in Euklidische Räume, *Monatshefte für Mathematik*, **59**, 3, pp. 217–229.

Boothby, W. (2002). *An Introduction to Differentiable Manifolds and Riemannian Geometry*, 2nd ed. (Academic Press).

Bourbaki, N. [1942](1998). *Algebra I: Chapters 1–3 (Elements of Mathematics)*, reprint ed. (Springer).

Bryant, V. (1971). Reducing Classical Axioms, *The Mathematical Gazette*, **55**, 391, pp. 38–40.

Caratheodory, C. (1909). Examination of the Foundations of Thermodynamics, *Mathematische Annalen*, **67**, pp. 355–386.

Cartan, É. (2002). *Riemannian Geometry in an Orthogonal Frame: From Lectures Delivered by Élie Cartan at the Sorbonne in 1926–27* (World Scientific).

Chang, H. S., Fu, M. C., Hu, J. and Marcus, S. I. (2005). An Adaptive Sampling Algorithm for Solving Markov Decision Processes, *Operations Research*, **53**, pp. 126–139.

Cohen, P. J. (1963). The Independence of the Continuum Hypothesis, *Proceedings of the National Academy of Sciences of the USA*, **50**, pp. 1143–1148.

Cohen, P. J. (1964). The Independence of the Continuum Hypothesis II, *Proceedings of the National Academy of Sciences of the USA*, **51**, pp. 105–110.

Coulom, R. (2006). Efficient Selectivity and Backup Operators in Monte-Carlo Tree Search, in *Proceedings of the 5th International Conference on Computers and Games*, pp. 72–83.

Dicke, R. H. (1957). Gravitation without a Principle of Equivalence, *Reviews of Modern Physics*, **29**, pp. 363–376.

Do Carmo, M. P. (1976). *Differential Geometry of Curves and Surfaces* (Prentice Hall).

Do Carmo, M. P. (1992). *Riemannian Geometry* (Birkhäuser).

Einstein, A. (1905). On the Electrodynamics of Moving Bodies, *Annalen der Physik*, **322** (10), pp. 891–921.

Einstein, A. (1916). The Foundation of the General Theory of Relativity, *Annalen der Physik*, **354**, pp. 769–822.

Einstein, A. (1961). *Relativity, the Special and General Theory* (Three Rivers Press).

Euclid, (1925). *The Thirteen Books of Euclid's Elements*, translated from the text of Heiberg with introduction and commentary (Cambridge University Press).

Friedman, M. (1983). *Foundations of Space-Time Theories: Relativistic Physics and Philosophy of Science* (Princeton University Press).

Gauss, C. F. (1827). *General Investigations of Curved Surfaces*, translated and edited by Morehead, J. C. and Hiltebeitel A. M., 1902, (Princeton).

Gergonne, J. D. (1818). Essai sur la Theorie des Definitions, *Annales de Mathématique Pure et Appliquée*, **9**, pp. 1–35.

Gibbs, J. W. (1884). *Elements of Vector Analysis: Arranged for the Use of Students in Physics* (Tuttle, Morehouse & Taylor).

Greub, W. H. (1967). *Multilinear Algebra* (Springer-Verlag).

Guo, H. and Mehrubeoglu, M. (2012). Analysis of Solar Panel Efficiency through Computation and Simulation, *International Journal of Modern Engineering*, **12**(2), pp. 45–52.

Guo, H. (2014). *Modern Mathematics and Applications in Computer Graphics and Vision* (World Scientific).

Guo, H. (2021). A New Paradox and the Reconciliation of Lorentz and Galilean Transformations, *Synthese*, https://doi.org/10.1007/s11229-021-03155-y (open access).

Hilbert, D. (1899). *Foundations of Geometry*, 2nd ed. 1971, (Open Court). Translated from the tenth German ed. First published in 1899.

Kline, M. (1972). *Mathematical Thought from Ancient to Modern Times* (Oxford University Press).

Kocsis, L. and Szepesvári, C. (2006). Bandit Based Monte-Carlo Planning, in *Proceedings of the 17th European Conference on Machine Learning (ECML 2006)*, pp. 282–293.

Kolb, R. (1996). *Blind Watchers of the Sky*, p. 299, (Addison-Wesley).

Levi-Civita, T. (1927). *The Absolute Differential Calculus (Calculus of Tensors)* (Blackie & Son).

Malament, D. (1977). Causal Theories of Time and the Conventionality of Simultaniety, *Noûs*, **11**, pp. 293–300.

Marcus, M. (1973). *Finite Dimensional Multilinear Algebra, Part I* (Marcel Dekker).

Mathematical Society of Japan (1993). *Encyclopedic Dictionary of Mathematics*, 2nd ed. (MIT Press).

Minkowski, H. (1908). The Fundamental Equations for Electromagnetic Processes in Moving Bodies. German Original: Die Grundgleichungen für die elektromagnetischen Vorgänge in bewegten Körpern, Nachrichten von der Gesellschaft der Wissenschaften zu Göttingen, Mathematisch-Physikalische Klasse, pp. 53–111. Presented in the session of December 21, 1907. Published in 1908.

Ohanian, H. (2009). *Einstein's Mistakes: The Human Failings of Genius* (Norton & Company).

O'Neill, B. (1997). *Elementary Differential Geometry*, 2nd ed. (Academic Press).

O'Neill, B. (1983). *Semi-Riemannian Geometry with Applications to Relativity* (Academic Press).

Peano, G. (1888). *Geometric Calculus: According to the Ausdehnungslehre of H. Grassmann*, translated by Kannenberg, L., 2000 (Birkhäuser). First published in 1888.

Poincaré, H. (1899). Des Fondements de la Géométrie, *Revue de Mëtaphysique et de Morale*, **7**, pp. 251–279.

Poincaré, H. (1905). *Science and Hypothesis* (Walter Scott Publishing).

Reich, K. (1994). Differential Geometry, in Grattan-Guinness, I. (ed.) *Companion Encyclopedia of the History and Philosophy of the Mathematical Sciences*, Vol. 1 (Routledge).

Reichenbach, H. [1924](1969). *Axiomatization of the Theory of Relativity* (University of California Press).

Ricci, G. (1892). Résumé de quelques travaux sur les systémes variables de fonctions, *Bulletin des Sciences Mathématiques*, **16**, pp. 167–189.

Ricci, G. and Levi-Civita, T. (1900). Methods of the Absolute Differential Calculus and Their Applications, *Mathematische Annalen*, **54**, pp. 125–201.

Rigby, J. F. and Wiegold, J. (1973). Independent Axioms for Vector Spaces, *The Mathematical Gazette*, **57**, 399, pp. 56–62.

Roman, S. (2005). *Advanced Linear Algebra*, 2nd ed. (Springer).

Rosen, N. (1940). General Relativity and Flat Space I, *Physical Review*, **57**, pp. 147–150.

Rosen, N. (1940). General Relativity and Flat Space II, *Physical Review*, **57**, pp. 150–153.

Selleri, F. (1996). Noninvariant One-Way Speed of Light, *Foundations of Physics*, **26**, pp. 641–664.

Selleri, F. (1997). Noninvariant One-Way Speed of Light and Locally Equivalent Reference Frames, *Foundations of Physics Letters*, **10**, pp. 73–83.

Silver, D. et al. (2016). Mastering the Game of Go with Deep Neural Networks and Tree Search, *Nature*, **529**, pp. 484–503.

Stein, S. (2010). *Mathematics: the Man-Made Universe*, 3rd revised ed. (Dover).

Struik, D. J. (1950). *Lectures on Classical Differential Geometry* (Addison-Wesley).

Sylvester, J. J. (1851). On the General Theory of Associated Algebraical Forms, *Cambridge and Dublin Mathematical Journal*, **4**, pp. 289–293.

Tarski, A. (1959). What Is Elementary Geometry?, in Henkin, L., Suppes, P. and Tarski, A. (ed.), *The Axiomatic Method: with special reference to geometry and physics. Proceedings of an International Symposium held at the Univ.*

of Calif., Berkeley, Dec. 26, 1957–Jan. 4, 1958, Studies in Logic and the Foundations of Mathematics, pp. 16–29 (North-Holland).

Voigt, W. (1898). *The Fundamental Physical Properties of the Crystals in an Elementary Representation* (Leipzig).

Weyl, H. (1918). Reine Infinitesimalgeometrie, *Mathematische Zeitschrift*, **2**, pp. 384–411.

Weyl, H. (1950). *Space-Time-Matter*, English translation by Brose, H. L. (Dover).

Whitney, H. (1938). Tensor Products of Abelian Groups, *Duke Mathematical Journal*, **4**, 3, pp. 495–528.

Wolf, R. (2005). *A Tour through Mathematical Logic* (Mathematical Association of America).

Index

9 789811 241017